Patricia Melin, Oscar Castillo

Hybrid Intelligent Systems for Pattern Recognition Using Soft Computing

Studies in Fuzziness and Soft Computing, Volume 172

Editor-in-chief
Prof. Janusz Kacprzyk
Systems Research Institute
Polish Academy of Sciences
ul. Newelska 6
01-447 Warsaw
Poland
E-mail: kacprzyk@ibspan.waw.pl

Patricia Melin
Oscar Castillo

Hybrid Intelligent Systems for Pattern Recognition Using Soft Computing

An Evolutionary Approach
for Neural Networks and Fuzzy Systems

 Springer

Professor Patricia Melin
Professor Oscar Castillo
Tijuana Institute of Technology
Department of Computer Science
P.O.Box 4207
Chula Vista, CA 91909, USA
E-mail: pmelin@tectijuana.mx
        ocastillo@tectijuana.mx

ISBN 978-3-642-06325-1   e-ISBN 978-3-540-32378-5
ISSN print edition:1434-9922
ISSN electronic edition:1860-0808

Springer is a part of Springer Science+Business Media
springeronline.com

Cover design: E. Kirchner, Springer Heidelberg
Printed on acid-free paper   62/3141/jl- 5 4 3 2 1 0

Preface

We describe in this book, new methods for intelligent pattern recognition using soft computing techniques. Soft Computing (SC) consists of several computing paradigms, including fuzzy logic, neural networks, and genetic algorithms, which can be used to produce powerful hybrid intelligent systems for pattern recognition. Hybrid intelligent systems that combine several SC techniques are needed due to the complexity and high dimensionality of pattern recognition problems. Hybrid intelligent systems can have different architectures, which have an impact on the efficiency and accuracy of pattern recognition systems, for this reason it is very important to optimize architecture design. The architectures can combine, in different ways, neural networks, fuzzy logic and genetic algorithms, to achieve the ultimate goal of pattern recognition. This book also shows results of the application of hybrid intelligent systems to real-world problems of face, fingerprint, and voice recognition.

As a prelude, we provide a brief overview of the existing methodologies in Soft Computing. We then describe our own approach in dealing with the problems in pattern recognition. Our particular point of view is that to really achieve intelligent pattern recognition in real-world applications we need to use SC techniques. As consequence, we will describe several real-world applications, in which the reader will be able to appreciate that the use of these techniques really helps in achieving the goals of intelligent pattern recognition. In these applications, we will always compare with the traditional approaches to make clear the advantages of using SC techniques.

This book is intended to be a major reference for scientists and engineers interested in applying new computational and mathematical tools to achieve intelligent pattern recognition. This book can also be used as a textbook or major reference for graduate courses like the following: soft computing, intelligent pattern recognition, computer vision, applied artificial intelligence, and similar ones. We consider that this book can also be used to get novel ideas for new lines of research, or to continue the lines of research proposed by the authors of the book.

In Chap. 1, we begin by giving a brief introduction to the main problems in achieving intelligent pattern recognition in real-world applications. We discuss the importance of the concept of intelligent pattern recognition. We motivate the need for using SC techniques for solving problems of face, fingerprint, and voice recognition. We also outline the real-world applications to be considered in the book.

We describe in Chap. 2 the main ideas underlying type-1 fuzzy logic, and the application of this powerful computational theory to the problems of modeling and pattern recognition. We discuss in some detail type-1 fuzzy set theory, fuzzy reasoning, and fuzzy inference systems. At the end, we also give some general guidelines for the process of fuzzy modeling. We illustrate these concepts with several examples that show the applicability of type-1 fuzzy logic. The importance of type-1 fuzzy logic as a basis for developing intelligent systems in pattern recognition has been recognized in several areas of application. For this reason, we consider this chapter essential to understand the new methods for intelligent pattern recognition that are described in subsequent chapters.

We describe in Chap. 3 the basic concepts, notation, and theory of type-2 fuzzy logic, and intuitionistic fuzzy logic, which are generalizations of type-1 fuzzy logic. Type-2 fuzzy logic enables the management of uncertainty in a more complete way. This is due to the fact that in type-2 membership functions we also consider that there is uncertainty in the form of the functions, unlike type-1 membership functions in which the functions are considered to be fixed and not uncertain. We describe type-2 fuzzy set theory, type-2 fuzzy reasoning, and type-2 fuzzy systems. We also give examples to illustrate these ideas to the reader of the book. Finally, we briefly describe the basic concepts and theory of intuitionistic fuzzy logic and illustrate their applicability with examples.

We describe in Chap. 4 the basic concepts, notation and the learning algorithms for supervised neural networks. We discuss in some detail feedforward neural networks, radial basis neural networks, and adaptive neuro-fuzzy inference systems. First, we give a brief review of the basic concepts of neural networks and the back-propagation learning algorithm. We then continue with a general description of radial basis neural networks. Finally, we end the chapter with a description of the adaptive neuro-fuzzy inference system (ANFIS) method and some examples of application. The importance of supervised neural networks as a computational tool to achieve "intelligence" for software systems has been well recognized in the literature of the area. For this reason, supervised neural networks have been applied for solving complex problems of modeling, identification, and pattern recognition.

We describe in Chap. 5 the basic concepts, notation and learning algorithms for unsupervised neural networks. This type of neural network only receives input data and not output data, unlike supervised neural networks, which receive input-output training data. We describe in some detail competitive neural networks, Kohonen sef-organizing maps, Learning Vector

Quantization (LVQ) neural networks, and Hopfield neural networks. We describe each of this type of neural networks and give examples to illustrate their applicability. Unsupervised neural networks are very important for classification, pattern recognition and clustering applications. For this reason, we consider this chapter very important for understanding some of the applications that are described in later chapters of the book.

We describe in Chap. 6 the basic concepts, theory and algorithms of modular and ensemble neural networks. We will also give particular attention to the problem of response integration, which is very important because response integration is responsible for combining all the outputs of the modules. Basically, a modular or ensemble neural network uses several monolithic neural networks to solve a specific problem. The basic idea is that combining the results of several simple neural networks we will achieve a better overall result in terms of accuracy and also learning can be done faster. For pattern recognition problems, which have great complexity and are defined over high dimensional spaces, modular neural networks are a great alternative for achieving the level of accuracy and efficiency needed for real-time applications. This chapter will serve as a basis for the modular architectures that will be proposed in later chapters for specific pattern recognition problems.

We describe in Chap. 7 the basic concepts and notation of genetic algorithms. We also describe the application of genetic algorithms for evolving neural networks, and fuzzy systems. Genetic algorithms are basic search methodologies that can be used for system optimization. Since genetic algorithms can be considered as a general-purpose optimization methodology, we can use it to find the model, which minimizes the fitting error for a specific data set. As genetic algorithms are based on the ideas of natural evolution, we can use this methodology to evolve a neural network or a fuzzy system for a particular application. The problem of finding the best architecture of a neural network is very important because there are no theoretical results on this, and in many cases we are forced to trial and error unless we use a genetic algorithm to automate this process. A similar thing occurs in finding out the optimal number of rules and membership functions of a fuzzy system for a particular application, here a genetic algorithm can also help us avoid time consuming trial and error.

We describe in Chap. 8 clustering with intelligent techniques, like fuzzy logic and neural networks. Cluster analysis is a technique for grouping data and finding structures in data. The most common application of clustering methods is to partition a data set into clusters or classes, where similar data are assigned to the same cluster whereas dissimilar data should belong to different clusters. In real-world applications there is very often no clear boundary between clusters so that fuzzy clustering is often a good alternative to use. Membership degrees between zero and one are used in fuzzy clustering instead of crisp assignments of the data to clusters. Pattern recognition techniques can be classified into two broad categories: *unsupervised* techniques and *supervised* techniques. An unsupervised technique does not use a given set of unclassified data points, whereas a supervised technique uses a data set with known classifications. These two types of techniques are complementary. For example, unsupervised clustering can be used to produce classification information needed by a supervised pattern recognition technique. In this chapter, we first give the basics of unsupervised clustering. The Fuzzy C-Means algorithm (FCM), which is the best known unsupervised fuzzy clustering algorithm is then described in detail. Supervised pattern recognition using fuzzy logic will also be mentioned. Finally, we describe the use of neural networks for unsupervised clustering and hybrid approaches.

We describe in Chap. 9 a new approach for face recognition using modular neural networks with a fuzzy logic method for response integration. We describe a new architecture for modular neural networks for achieving pattern recognition in the particular case of human faces. Also, the method for achieving response integration is based on the fuzzy Sugeno integral. Response integration is required to combine the outputs of all the modules in the modular network. We have applied the new approach for face recognition with a real database of faces from students and professors of our institution. Recognition rates with the modular approach were compared against the monolithic single neural network approach, to measure the improvement. The results of the new modular neural network approach gives excellent performance overall and also in comparison with the monolithic approach. The chapter is divided as follows: first we give a brief introduction to the problem of face recognition, second we describe the proposed architecture for achieving face recognition, third, we describe the fuzzy method for response integration, and finally we show a summary of the results and conclusions.

We describe in Chap. 10 a new approach for fingerprint recognition using modular neural networks with a fuzzy logic method for response integration. We describe a new architecture for modular neural networks for achieving pattern recognition in the particular case of human fingerprints. Also, the method for achieving response integration is based on the fuzzy Sugeno integral. Response integration is required to combine the outputs of all the modules in the modular network. We have applied the new approach for fingerprint recognition with a real database of fingerprints obtained from students of our institution.

We describe in Chap. 11 the use of neural networks, fuzzy logic and genetic algorithms for voice recognition. In particular, we consider the case of speaker recognition by analyzing the sound signals with the help of intelligent techniques, such as the neural networks and fuzzy systems. We use the neural networks for analyzing the sound signal of an unknown speaker, and after this first step, a set of type-2 fuzzy rules is used for decision making. We need to use fuzzy logic due to the uncertainty of the decision process. We also use genetic algorithms to optimize the architecture of the neural networks. We illustrate our approach with a sample of sound signals from real speakers in our institution.

We describe in Chap. 12 a new approach for human recognition using as information the face, fingerprint, and voice of a person. We have described in the previous chapters the use of intelligent techniques for achieving face recognition, fingerprint recognition, and voice identification. Now in this chapter we are considering the integration of these three biometric measures to improve the accuracy of human recognition. The new approach will integrate the information from three main modules, one for each of the three biometric measures. The new approach consists in a modular architecture that contains three basic modules: face, fingerprint, and voice. The final decision is based on the results of the three modules and uses fuzzy logic to take into account the uncertainty of the outputs of the modules.

We end this preface of the book by giving thanks to all the people who have help or encourage us during the writing of this book. First of all, we would like to thank our colleague and friend Prof. Janusz Kacprzyk for always supporting our work, and for motivating us to write our research work. We would also like to thank our colleagues working in Soft Computing, which are too many to mention each by their name. Of course, we need to thank our supporting agencies, CONACYT and COSNET, in our country for their help during this project. We have to thank our institution, Tijuana Institute of Technology, for always supporting our projects. Finally, we thank our families for their continuous support during the time that we spend in this project.

Mexico   *Patricia Melin*
January 2005   *Oscar Castillo*

Contents

1

Introduction to Pattern Recognition with Intelligent Systems

We describe in this book, new methods for intelligent pattern recognition using soft computing techniques. Soft Computing (SC) consists of several computing paradigms, including fuzzy logic, neural networks, and genetic algorithms, which can be used to create powerful hybrid intelligent systems. Combining SC techniques, we can build powerful hybrid intelligent systems that can use the advantages that each technique offers. We consider in this book "intelligent pattern recognition" as the use of SC techniques to solve pattern recognition problems in real-world applications. We consider in particular the problems of face, fingerprint and voice recognition. We also consider the problem of recognizing a person by integrating the information given by the face, fingerprint and voice of the person.

As a prelude, we provide a brief overview of the existing methodologies for solving pattern recognition problems. We then describe our own approach in dealing with these problems. Our particular point of view is that face recognition, fingerprint recognition and voice identification are problems that can not be considered apart because they are intrinsically related in real-world applications. We show in this book that face recognition can be achieved by using modular neural networks and fuzzy logic. Genetic algorithms can also be use to optimize the architecture of the face recognition system. Fingerprint recognition can also be achieved by applying modular neural networks and fuzzy logic in a similar way as in the method for face recognition. Finally, voice recognition can be achieved by applying neural networks, fuzzy logic, and genetic algorithms. We will illustrate in this book each of these recognition problems and its solutions in real world situations. In each application of the SC techniques to solve a real-world pattern recognition problem, we show that the intelligent approach proves to be more efficient and accurate that traditional approaches.

Traditionally, pattern recognition problems mentioned above, have been solved by using classical statistical methods and models, which lack, in some cases, the accuracy and efficiency needed in real-world applications. Traditional methods include the use of statistical models and simple information

Patricia Melin and Oscar Castillo: *Hybrid Intelligent Systems for Pattern Recognition Using Soft Computing*, StudFuzz **172**, 1–5 (2005)
www.springerlink.com

systems. We instead, consider more general modeling methods, which include fuzzy logic and neural networks. We also use genetic algorithms for the optimization of the fuzzy systems and neural networks. A proper combination of these methodologies will result in a hybrid intelligent system that will solve efficiently and accurately a specific pattern recognition problem.

1.1 Soft Computing Techniques for Pattern Recognition

Fuzzy logic is an area of soft computing that enables a computer system to reason with uncertainty (Castillo & Melin, 2001). A fuzzy inference system consists of a set of if-then rules defined over fuzzy sets. Fuzzy sets generalize the concept of a traditional set by allowing the membership degree to be any value between 0 and 1 (Zadeh, 1965). This corresponds, in the real world, to many situations where it is difficult to decide in an unambiguous manner if something belongs or not to a specific class. Fuzzy expert systems, for example, have been applied with some success to problems of decision, control, diagnosis and classification, just because they can manage the complex expert reasoning involved in these areas of application. The main disadvantage of fuzzy systems is that they can't adapt to changing situations. For this reason, it is a good idea to combine fuzzy logic with neural networks or genetic algorithms, because either one of these last two methodologies could give adaptability to the fuzzy system (Melin & Castillo, 2002). On the other hand, the knowledge that is used to build these fuzzy rules is uncertain. Such uncertainty leads to rules whose antecedents or consequents are uncertain, which translates into uncertain antecedent or consequent membership functions (Karnik & Mendel 1998). Type-1 fuzzy systems, like the ones mentioned above, whose membership functions are type-1 fuzzy sets, are unable to directly handle such uncertainties. We also describe in this book, type-2 fuzzy systems, in which the antecedent or consequent membership functions are type-2 fuzzy sets. Such sets are fuzzy sets whose membership grades themselves are type-1 fuzzy sets; they are very useful in circumstances where it is difficult to determine an exact membership function for a fuzzy set. Another way to handle this higher degree of uncertainty is to use intuitionistic fuzzy logic (Atanassov, 1999), which can also be considered as a generalization of type-1 fuzzy logic. In intuitionistic fuzzy logic the uncertainty in describing fuzzy sets is modeled by using at the same time the membership function and the non-membership function of a set (assuming that they are not complementary).

Neural networks are computational models with learning (or adaptive) characteristics that model the human brain (Jang, Sun & Mizutani, 1997). Generally speaking, biological natural neural networks consist of neurons and connections between them, and this is modeled by a graph with nodes and arcs to form the computational neural network. This graph along with a computational algorithm to specify the learning capabilities of the system is what makes the neural network a powerful methodology to simulate intelligent or expert behavior (Miller, Sutton & Werbos, 1995). Neural networks can be classified in supervised and unsupervised. The main difference is that in the case of the supervised neural networks the learning algorithm uses input-output training data to model the dynamic system, on the other hand, in the case of unsupervised neural networks only the input data is given. In the case of an unsupervised network, the input data is used to make representative clusters of all the data. It has been shown, that neural networks are universal approximators, in the sense that they can model any general function to a specified accuracy and for this reason neural networks have been applied to problems of system identification, control, diagnosis, time series prediction, and pattern recognition. We also describe the basic concepts, theory and algorithms of modular and ensemble neural networks. We will also give particular attention to the problem of response integration, which is very important because response integration is responsible for combining all the outputs of the modules. Basically, a modular or ensemble neural network uses several monolithic neural networks to solve a specific problem. The basic idea is that combining the results of several simple neural networks we will achieve a better overall result in terms of accuracy and also learning can be done faster. For pattern recognition problems, which have great complexity and are defined over high dimensional spaces, modular neural networks are a great alternative for achieving the level of accuracy and efficiency needed for real-time applications.

Genetic algorithms and evolutionary methods are optimization methodologies based on principles of nature (Jang, Sun & Mizutani, 1997). Both methodologies can also be viewed as searching algorithms because they explore a space using heuristics inspired by nature. Genetic algorithms are based on the ideas of evolution and the biological process that occur at the DNA level. Basically, a genetic algorithm uses a population of individuals, which are modified by using genetic operators in such a way as to eventually obtain the fittest individual (Man, Tang & Kwong, 1999). Any optimization problem has to be represented by using chromosomes, which are a codified representation of the real values of the variables in the problem (Mitchell, 1998). Both, genetic algorithms and evolutionary methods can be used to optimize a general objective function. As genetic algorithms are based on the ideas of natural evolution, we can use this methodology to evolve a neural network or a fuzzy system for a particular application. The problem of finding the best architecture of a neural network is very important because there are no theoretical results on this, and in many cases we are forced to trial and error unless we use a genetic algorithm to automate this process. A similar thing occurs in finding out the optimal number of rules and membership functions of a fuzzy system for a particular application, here a genetic algorithm can also help us avoid time consuming trial and error. In this book, we use genetic algorithms to optimize the architecture of pattern recognition systems.

1.2 Pattern Recognition Applications

We consider in this book the pattern recognition problems of face, fingerprint and voice recognition. Although the methods that are described can also be used for other type of pattern recognition problems, we concentrate here on the problems mentioned above. Pattern recognition, in many cases, requires that we perform a previous step of clustering the data to facilitate the use of the recognition methods. For this reason, we describe in detail methods of clustering data. We also describe the new hybrid intelligent approaches for achieving pattern recognition of the problems mentioned above.

We describe in this book clustering with intelligent techniques, like fuzzy logic and neural networks. Cluster analysis is a technique for grouping data and finding structures in data. The most common application of clustering methods is to partition a data set into clusters or classes, where similar data are assigned to the same cluster whereas dissimilar data should belong to different clusters. In real-world applications there is very often no clear boundary between clusters so that fuzzy clustering is often a good alternative to use. Membership degrees between zero and one are used in fuzzy clustering instead of crisp assignments of the data to clusters. Pattern recognition techniques can be classified into two broad categories: *unsupervised* techniques and *supervised* techniques. An unsupervised technique does not use a given set of unclassified data points, whereas a supervised technique uses a data set with known classifications. These two types of techniques are complementary. For example, unsupervised clustering can be used to produce classification information needed by a supervised pattern recognition technique. The Fuzzy C-Means algorithm (FCM), which is the best known unsupervised fuzzy clustering algorithm, is then described in detail. Supervised pattern recognition using fuzzy logic will also be mentioned. Finally, we describe the use of neural networks for unsupervised clustering and hybrid approaches.

We describe in this book a new approach for face recognition using modular neural networks with a fuzzy logic method for response integration. We describe a new architecture for modular neural networks for achieving pattern recognition in the particular case of human faces. Also, the method for achieving response integration is based on the fuzzy Sugeno integral. Response integration is required to combine the outputs of all the modules in the modular network. We have applied the new approach for face recognition with a real database of faces from students and professors of our institution. Recognition rates with the modular approach were compared against the monolithic single neural network approach, to measure the improvement. The results of the new modular neural network approach gives excellent performance overall and also in comparison with the monolithic approach.

We describe in this book a new approach for fingerprint recognition using modular neural networks with a fuzzy logic method for response integration. We describe a new architecture for modular neural networks for achieving pattern recognition in the particular case of human fingerprints. Also, the method for achieving response integration is based on the fuzzy Sugeno integral. Response integration is required to combine the outputs of all the modules in the modular network. We have applied the new approach for fingerprint recognition with a real database of fingerprints obtained from students of our institution.

We also describe in this book the use of neural networks, fuzzy logic and genetic algorithms for voice recognition. In particular, we consider the case of speaker recognition by analyzing the sound signals with the help of intelligent techniques, such as the neural networks and fuzzy systems. We use the neural networks for analyzing the sound signal of an unknown speaker, and after this first step, a set of type-2 fuzzy rules is used for decision making. We need to use fuzzy logic due to the uncertainty of the decision process. We also use genetic algorithms to optimize the architecture of the neural networks. We illustrate our approach with a sample of sound signals from real speakers in our institution.

Finally, we describe in this book our new approach for human recognition using as information the face, fingerprint, and voice of a person. We have described above the use of intelligent techniques for achieving face recognition, fingerprint recognition, and voice identification. Now we can consider the integration of these three biometric measures to improve the accuracy of human recognition. The new approach will integrate the information from three main modules, one for each of the three biometric measures. The new approach consists in a modular architecture that contains three basic modules: face, fingerprint, and voice. The final decision is based on the results of the three modules and uses fuzzy logic to take into account the uncertainty of the outputs of the modules.

2

Type-1 Fuzzy Logic

This chapter introduces the basic concepts, notation, and basic operations for the type-1 fuzzy sets that will be needed in the following chapters. Type-2 fuzzy sets as well as their operations will be discussed in the next chapter. For this reason, in this chapter we will focus only on type-1 fuzzy logic. Since research on fuzzy set theory has been underway for over 30 years now, it is practically impossible to cover all aspects of current developments in this area. Therefore, the main goal of this chapter is to provide an introduction to and a summary of the basic concepts and operations that are relevant to the study of type-1 fuzzy sets. We also introduce in this chapter the definition of linguistic variables and linguistic values and explain how to use them in type-1 fuzzy rules, which are an efficient tool for quantitative modeling of words or sentences in a natural or artificial language. By interpreting fuzzy rules as fuzzy relations, we describe different schemes of fuzzy reasoning, where inference procedures based on the concept of the compositional rule of inference are used to derive conclusions from a set of fuzzy rules and known facts. Fuzzy rules and fuzzy reasoning are the basic components of fuzzy inference systems, which are the most important modeling tool, based on fuzzy set theory.

The "fuzzy inference system" is a popular computing framework based on the concepts of fuzzy set theory, fuzzy if-then rules, and fuzzy reasoning (Jang, Sun & Mizutani, 1997). It has found successful applications in a wide variety of fields, such as automatic control, data classification, decision analysis, expert systems, time series prediction, robotics, and pattern recognition (Jamshidi, 1997). Because of its multidisciplinary nature, the fuzzy inference system is known by numerous other names, such as "fuzzy expert system" (Kandel, 1992), "fuzzy model" (Sugeno & Kang, 1988), "fuzzy associative memory" (Kosko, 1992), and simply "fuzzy system".

The basic structure of a type-1 fuzzy inference system consists of three conceptual components: a "rule base", which contains a selection of fuzzy rules; a "data base" (or "dictionary"), which defines the membership functions used in the fuzzy rules; and a "reasoning mechanism", which performs the inference procedure upon the rules and given facts to derive a reasonable

Patricia Melin and Oscar Castillo: *Hybrid Intelligent Systems for Pattern Recognition Using Soft Computing*, StudFuzz **172**, 7–32 (2005)
www.springerlink.com

output or conclusion. In general, we can say that a fuzzy inference system implements a non-linear mapping from its input space to output space. This mapping is accomplished by a number of fuzzy if-then rules, each of which describes the local behavior of the mapping. In particular, the antecedent of a rule defines a fuzzy region in the input space, while the consequent specifies the output in the fuzzy region.

We will describe in the following chapter a new area in fuzzy logic, which studies type-2 fuzzy sets and type-2 fuzzy systems. Basically, a type-2 fuzzy set is a set in which we also have uncertainty about the membership function. Since we are dealing with uncertainty for the conventional fuzzy sets (which are called type-1 fuzzy sets here) we can achieve a higher degree of approximation in modeling real world problems. Of course, type-2 fuzzy systems consist of fuzzy if-then rules, which contain type-2 fuzzy sets. We can say that type-2 fuzzy logic is a generalization of conventional fuzzy logic (type-1) in the sense that uncertainty is not only limited to the linguistic variables but also is present in the definition of the membership functions.

In what follows, we shall first introduce the basic concepts of fuzzy sets, and fuzzy reasoning. Then we will introduce and compare the three types of fuzzy inference systems that have been employed in various applications. Finally, we will address briefly the features and problems of fuzzy modeling, which is concerned with the construction of fuzzy inference systems for modeling a given target system. In this chapter, we will assume that all fuzzy sets, fuzzy rules and operations are of type-1 category, unless otherwise specified.

2.1 Type-1 Fuzzy Set Theory

Let X be a space of objects and x be a generic element of X. A classical set $A, A \subseteq X$, is defined by a collection of elements or objects $x \in X$, such that each x can either belong or not belong to the set A. By defining a "characteristic function" for each element $x \in X$, we can represent a classical set A by a set of order pairs $(x, 0)$ or $(x, 1)$, which indicates $x \notin A$ or $x \in A$, respectively.

Unlike the aforementioned conventional set, a fuzzy set (Zadeh, 1965) expresses the degree to which an element belong to a set. Hence the characteristic function of a fuzzy set is allowed to have values between 0 and 1, which denotes the degree of membership of an element in a given set.

Definition 2.1. *Fuzzy sets and membership functions*
If X is a collection of objects denoted generically by x, then a "fuzzy set" A in X is defined as a set of ordered pairs:

$$A = \{(x, \mu_A(x)) \mid x \in X\}. \tag{2.1}$$

where $\mu_A(x)$ is called "membership function" (or MF for short) for the fuzzy set A. The MF maps each element of X to a membership grade (or membership value) between 0 and 1.

Obviously, the definition of a fuzzy set is a simple extension of the definition of a classical set in which the characteristic function is permitted to have any values between 0 and 1. If the values of the membership function $\mu_A(x)$ is restricted to either 0 or 1, then A is reduced to a classical set and $\mu_A(x)$ is the characteristic function of A. This can be seen with the following example.

Example 2.1. Fuzzy set with a discrete universe of discourse X
Let $X = \{$Tijuana, Acapulco, Cancun$\}$ be the set of cities one may choose to organize a conference in. The fuzzy set $A = $ "desirable city to organize a conference in" may be described as follows:

$$A = \{(\text{Tijuana}, 0.5), (\text{Acapulco}, 0.7), (\text{Cancun}, 0.9)\}$$

In this case, the universe of discourse X is discrete – in this example, three cities in Mexico. Of course, the membership grades listed above are quite subjective; anyone can come up with three different values according to his or her preference.

A fuzzy set is uniquely specified by its membership function. To describe membership functions more specifically, we shall define the nomenclature used in the literature (Jang, Sun & Mizutani, 1997).

Definition 2.2. *Support*
The "support" of a fuzzy set A is the set of all points x in X such that $\mu_A(x) > 0$:
$$\text{support}(A) = \{x \,|\, \mu_A(x) > 0\} \,. \tag{2.2}$$

Definition 2.3. *Core*
The "core" of a fuzzy set is the set of all points x in X such that $\mu_A(x) = 1$:
$$\text{core}(A) = \{x \,|\, \mu_A(x) = 1\} \,. \tag{2.3}$$

Definition 2.4. *Normality*
A fuzzy set A is "normal" if its core is nonempty. In other words, we can always find a point $x \in X$ such that $\mu_A(x) = 1$.

Definition 2.5. *Crossover points*
A "crossover point" of a fuzzy set A is a point $x \in X$ at which $\mu_A(x) = 0.5$:
$$\text{crossover}(A) = \{x \,|\, \mu_A(x) = 0.5\} \,. \tag{2.4}$$

Definition 2.6. *Fuzzy singleton*
A fuzzy set whose support is a single point in X with $\mu_A(x) = 1$ is called a "fuzzy singleton".

Definition 2.7. α-cut, strong α-cut
The "α-cut" or "α-level set" of a fuzzy set A is a crisp set defined by

$$A_\alpha = \{x \mid \mu_A(x) \geq \alpha\} . \tag{2.5}$$

"Strong α-cut" or "strong αt-level set" are defined similarly:

$$A'_\alpha = \{x \mid \mu_A(x) > \alpha\} . \tag{2.6}$$

Using the notation for a level set, we can express the support and core of a fuzzy set A as

$$\text{support}(A) = A'_0$$

and

$$\text{core}(A) = A_1$$

respectively.

Corresponding to the ordinary set operations of union, intersection and complement, fuzzy sets have similar operations, which were initially defined in Zadeh's seminal paper (Zadeh, 1965). Before introducing these three fuzzy set operations, first we shall define the notion of containment, which plays a central role in both ordinary and fuzzy sets. This definition of containment is, of course, a natural extension of the case for ordinary sets.

Definition 2.8. *Containment*
The fuzzy set A is "contained" in fuzzy set B (or, equivalently, A is a "subset" of B) if and only if $\mu_A(x) \leq \mu_B(x)$ for all x. Mathematically,

$$A \subseteq B \Leftrightarrow \mu_A(x) \leq \mu_B(x) . \tag{2.7}$$

Definition 2.9. *Union*
The "union" of two fuzzy sets A and B is a fuzzy set C, written as $C = A \cup B$ or $C = A$ OR B, whose MF is related to those of A and B by

$$\mu_C(x) = \max(\mu_A(x), \mu_B(x)) = \mu_A(x) \vee \mu_B(x) . \tag{2.8}$$

Definition 2.10. *Intersection*
The "intersection" of two fuzzy sets A and B is a fuzzy set C, written as $C = A \cap B$ or $C = A$ AND B, whose MF is related to those of A and B by

$$\mu_C(x) = \min(\mu_A(x), \mu_B(x)) = \mu_A(x) \wedge \mu_B(x) . \tag{2.9}$$

Definition 2.11. *Complement or Negation*
The "complement" of a fuzzy set A, denoted by $A(\rceil A, NOT\ A)$, is defined as

$$\mu_A(x) = 1 - \mu_A(x) . \tag{2.10}$$

As mentioned earlier, a fuzzy set is completely characterized by its MF. Since most fuzzy sets in use have a universe of discourse X consisting of the real line R, it would be impractical to list all the pairs defining a membership function. A more convenient and concise way to define a MF is to express it as a mathematical formula. First we define several classes of parameterized MFs of one dimension.

Definition 2.12. *Triangular MFs*
A "triangular MF" is specified by three parameters $\{a, b, c\}$ as follows:

$$y = \text{triangle}(x; a, b, c) = \begin{cases} 0, & x \le a. \\ (x-a)/(b-a), & a \le x \le b. \\ (c-x)/(c-b), & b \le x \le c. \\ 0, & c \le x. \end{cases} \quad (2.11)$$

The parameters $\{a, b, c\}$ (with $a < b < c$) determine the x coordinates of the three corners of the underlying triangular MF. Figure 2.1(a) illustrates a triangular MF defined by triangle$(x; 10, 20, 40)$.

Definition 2.13. *Trapezoidal MFs*
A "trapezoidal MF" is specified by four parameters $\{a, b, c, d\}$ as follows:

$$\text{trapezoid}(x; a, b, c, d) = \begin{cases} 0, & x \le a. \\ (x-a)/(b-a), & a \le x \le b. \\ 1, & b \le x \le c. \\ (d-x)/(d-c), & c \le x \le d. \\ 0, & d \le x. \end{cases} \quad (2.12)$$

The parameters $\{a, b, c, d\}$ (with $a < b \le c < d$) determine the x coordinates of the four corners of the underlying trapezoidal MF. Figure 2.1(b) illustrates a trapezoidal MF defined by trapezoid$(x; 10, 20\ 40, 75)$.

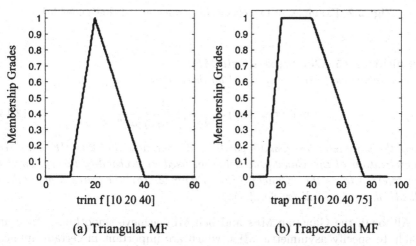

(a) Triangular MF   (b) Trapezoidal MF

Fig. 2.1. Examples of two types of parameterized MFs

Due to their simple formulas and computational efficiency, both triangular MFs and trapezoidal MFs have been used extensively, especially in real-time implementations. However, since the MFs are composed of straight line segments, they are not smooth at the corner points specified by the parameters. In the following we introduce other types of MFs defined by smooth and nonlinear functions.

Definition 2.14. *Gaussian MFs*
A "Gaussian MF" is specified by two parameters $\{c, \sigma\}$:

$$\text{gaussian} (x; c, \sigma) = e^{-\frac{1}{2}\frac{(x-c)^2}{\sigma}} . \tag{2.13}$$

A "Gaussian" MF is determined completely by c and σ; c represents the MFs center and σ determines the MFs width. Figure 2.2(a) plots a Gaussian MF defined by gaussian $(x; 50, 20)$.

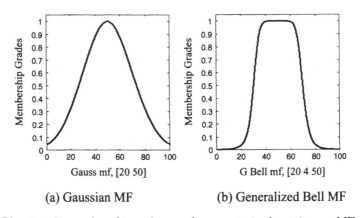

(a) Gaussian MF   (b) Generalized Bell MF

Fig. 2.2. Examples of two classes of parameterized continuous MFs

Definition 2.15. *Generalized bell MFs*
A "generalized bell MF" is specified by three parameters $\{a, b, c\}$:

$$\text{bell} (x; a, b, c) = \frac{1}{1 + |(x - c)/a|^{2b}} \tag{2.14}$$

where the parameter b is usually positive. We can note that this MF is a direct generalization of the Cauchy distribution used in probability theory, so it is also referred to as the "Cauchy MF". Figure 2.2(b) illustrates a generalized bell MF defined by bell $(x; 20, 4, 50)$.

Although the Gaussian MFs and bell MFs achieve smoothness, they are unable to specify asymmetric MFs, which are important in certain applications. Next we define the sigmoidal MF, which is either open left or right.

Definition 2.16. *Sigmoidal MFs*

A "Sigmoidal MF" is defined by the following equation:

$$\text{sig}(x;\ a,c) = \frac{1}{1 + \exp[-a(x-c)]} \tag{2.15}$$

where a controls the slope at the crossover point $x = c$.

Depending on the sign of the parameter "a", a sigmoidal MF is inherently open right or left and thus is appropriate for representing concepts such as "very large" or "very negative". Figure 2.3 shows two sigmoidal functions $y_1 = \text{sig}(x;\ 1,\ -5)$ and $y_2 = \text{sig}(x;\ -2,\ 5)$.

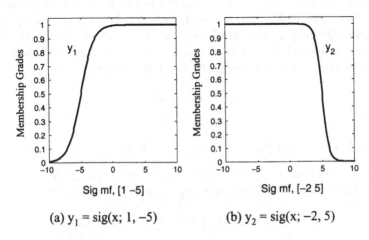

(a) $y_1 = \text{sig}(x;\ 1, -5)$ (b) $y_2 = \text{sig}(x;\ -2, 5)$

Fig. 2.3. Two sigmoidal functions y_1 and y_2

2.2 Fuzzy Rules and Fuzzy Reasoning

In this section we introduce the concepts of the extension principle and fuzzy relations, which extend the notions of fuzzy sets introduced previously. Then we give the definition of linguistic variables and linguistic values and show how to use them in fuzzy rules. By interpreting fuzzy rules as fuzzy relations, we describe different schemes of fuzzy reasoning. Fuzzy rules and fuzzy reasoning are the backbone of fuzzy inference systems, which are the most important modeling tool based on fuzzy set theory.

2.2.1 Fuzzy Relations

The "extension principle" is a basic concept of fuzzy set theory that provides a general procedure for extending crisp domains of mathematical expressions to fuzzy domains. This procedure generalizes a common one-to-one mapping of a function f to a mapping between fuzzy sets. More specifically, lets assume that f is a function from X to Y and A is a fuzzy set on X defined as

$$A = \mu_A(x_1)/x_1 + \mu_A(x_2)/x_2 + \cdots + \mu_A(x_n)/x_n$$

Then the extension principle states that the image of fuzzy set A under the mapping f can be expressed as a fuzzy set B,

$$B = f(A) = \mu_A(x_1)/y_1 + \mu_A(x_2)/y_2 + \cdots + \mu_A(x_n)/y_n$$

where $y_i = f(x_i)$, $i = 1, \ldots, n$. In other words, the fuzzy set B can be defined through the values of f in $x_1, x_2, \ldots, x_n$. If f is a many-to-one mapping, then there exists $x_1, x_2 \in X$, $x_1 \neq x_2$, such that $f(x_1) = f(x_2) = y^*$, $y^* \in Y$. In this case, the membership grade of B at $y = y^*$ is the maximum of the membership grades of A at $x = x_1$ and $x = x_2$, since $f(x) = y^*$ may result from $x = x_1$ or $x = x_2$. More generally speaking, we have

$$\mu_B(y) = \max_{x = f^{-1}(y)} \mu_A(x) \,.$$

A simple example of this concept is shown below.

Example 2.2. Application of the extension principle to fuzzy sets
Lets suppose we have the following fuzzy set with discrete universe

$$A = 0.2/-2 + 0.5/-1 + 0.7/0 + 0.9/1 + 0.4/2$$

and lets suppose that we have the following mapping

$$y = x^2 + 1 \,.$$

After applying the extension principle, we have the following result

$$B = 0.2/5 + 0.5/2 + 0.7/1 + 0.9/2 + 0.4/5$$
$$B = 0.7/1 + (0.2 \vee 0.4)/5 + (0.5 \vee 0.9)/2$$
$$B = 0.7/1 + 0.4/5 + 0.9/2 \,,$$

where $\vee$ represents "max".

Binary fuzzy relations are fuzzy sets in $X \times Y$ which map each element in $X \times Y$ to a membership grade between 0 and 1. In particular, unary fuzzy relations are fuzzy sets with one-dimensional MFs; binary fuzzy relations are fuzzy sets with two-dimensional MFs, and so on. Here we will restrict our attention to binary fuzzy relations. A generalization to n-ary fuzzy relations is not so difficult.

Definition 2.17. *Binary fuzzy relation*
Let X and Y be two universes of discourse. Then

$$\Re = \{((x, y), \mu_\Re(x, y))|(x, y) \in X \times Y\} \qquad (2.16)$$

is a binary fuzzy relation in $X \times Y$.

Example 2.3. Binary fuzzy relations
Let $X = \{1, 2, 3\}$ and $Y = \{1, 2, 3, 4, 5\}$ and $\Re$ = "y is slightly greater than x". The MF of the fuzzy relation $\Re$ can be defined (subjectively) as

$$\mu_\Re(x, y) = \begin{cases} (y - x)/(y + x), & \text{if } y > x \, . \\ 0, & \text{if } y \leq x \, . \end{cases} \qquad (2.17)$$

This fuzzy relation $\Re$ can be expressed as a relation matrix in the following form:

$$\Re = \begin{pmatrix} 0 & 0.333 & 0.500 & 0.600 & 0.666 \\ 0 & 0 & 0.200 & 0.333 & 0.428 \\ 0 & 0 & 0 & 0.142 & 0.250 \end{pmatrix}$$

where the element at row i and column j is equal to the membership grade between the ith element of X and jth element of Y.

Other common examples of binary fuzzy relations are the following:

- x is similar to y (x and y are objects)
- x depends on y (x and y are events)
- If x is big, then y is small (x is an observed reading and y is the corresponding action)

The last example, "If x is A, then y is B", is used repeatedly in fuzzy systems. We will explore fuzzy relations of this type in the following section.

Fuzzy relations in different product spaces can be combined through a composition operation. Different composition operations have been proposed for fuzzy relations; the best known is the max-min composition proposed by Zadeh, in 1965.

Definition 2.18. *Max-min composition*
Let $\Re_1$ and $\Re_2$ be two fuzzy relations defined on $X \times Y$ and $Y \times Z$, respectively. The "max-min composition" of $\Re_1$ and $\Re_2$ is a fuzzy set defined by

$$\Re_1 \circ \Re_2 = \{[(x, z), \max_y \min(\mu_{\Re_1}(x, y), \mu_{\Re_2}(y, z))]|x \in X, y \in Y, z \in Z\}$$
$$(2.18)$$

When $\Re_1$ and $\Re_2$ are expressed as relation matrices, the calculation of the composition $\Re_1 \circ \Re_2$ is almost the same as matrix multiplication, except that $\times$ and $+$ are replaced by the "min" and "max" operations, respectively. For this reason, the max-min composition is also called the "max-min product".

2.2.2 Fuzzy Rules

As was pointed out by Zadeh in his work on this area (Zadeh, 1973), conventional techniques for system analysis are intrinsically unsuited for dealing with humanistic systems, whose behavior is strongly influenced by human judgment, perception, and emotions. This is a manifestation of what might be called the "principle of incompatibility": "As the complexity of a system increases, our ability to make precise and yet significant statements about its behavior diminishes until a threshold is reached beyond which precision and significance become almost mutually exclusive characteristics" (Zadeh, 1973). It was because of this belief that Zadeh proposed the concept of linguistic variables (Zadeh, 1971) as an alternative approach to modeling human thinking.

Definition 2.19. *Linguistic variables*
A "Linguistic variable" is characterized by a quintuple $(x, T(x), X, G, M)$ in which x is the name of the variable; $T(x)$ is the "term set" of x-that is, the set of its "linguistic values" or "linguistic terms"; X is the universe of discourse, G is a "syntactic rule" which generates the terms in $T(x)$; and M is a "semantic rule" which associates with each linguistic value A its meaning $M(A)$, where $M(A)$ denotes a fuzzy set in X.

Definition 2.20. *Concentration and dilation of linguistic values*
Let A be a linguistic value characterized by a fuzzy set membership function $\mu_A(.)$. Then A^k is interpreted as a modified version of the original linguistic value expressed as

$$A^k = \int_x [\mu_A(x)]^k /x \ . \tag{2.19}$$

In particular, the operation of "concentration" is defined as

$$\mathrm{CON}(A) = A^2 \ , \tag{2.20}$$

while that of "dilation" is expressed by

$$\mathrm{DIL}(A) = A^{0.5} \ . \tag{2.21}$$

Conventionally, we take $\mathrm{CON}(A)$ and $\mathrm{DIL}(A)$ to be the results of applying the hedges "very" and "more or less", respectively, to the linguistic term A. However, other consistent definitions for these linguistic hedges are possible and well justified for various applications.

Following the definitions given before, we can interpret the negation operator NOT and the connectives AND and OR as

$$\mathrm{NOT}(A) = \rceil A = \int_x [1 - \mu_A(x)]/x \ ,$$

$$A \ \mathrm{AND} \ B = A \cap B = \int_x [\mu_A(x) \wedge \mu_B(x)]/x \ , \tag{2.22}$$

$$A \ \mathrm{OR} \ B = A \cup B = \int_x [\mu_A(x) \vee \mu_B(x)]/x \ .$$

respectively, where A and B are two linguistic values whose meanings are defined by $\mu_A(.)$ and $\mu_B(.)$.

Definition 2.21. *Fuzzy If-Then Rules*
A "fuzzy if-then rule" (also known as "fuzzy rule", "fuzzy implication", or "fuzzy conditional statement") assumes the form

$$\text{if } x \text{ is } A \text{ then } y \text{ is } B ,\qquad (2.23)$$

where A and B are linguistic values defined by fuzzy sets on universes of discourse X and Y, respectively. Often "x is A" is called "antecedent" or "premise", while "y is B" is called the "consequence" or "conclusion".

Examples of fuzzy if-then rules are widespread in our daily linguistic expressions, such as the following:

- If pressure is high, then volume is small.
- If the road is slippery, then driving is dangerous.
- If the speed is high, then apply the brake a little.

Before we can employ fuzzy if-then rules to model and analyze a system, first we have to formalize what is meant by the expression "if x is A then y is B", which is sometimes abbreviated as $A \rightarrow B$. In essence, the expression describes a relation between two variables x and y; this suggests that a fuzzy if-then rule is defined as a binary fuzzy relation R on the product space $X \times Y$. Generally speaking, there are two ways to interpret the fuzzy rule $A \rightarrow B$. If we interpret $A \rightarrow B$ as A "coupled with" B then

$$R = A \rightarrow B = A \times B = \int_{x \times y} \mu_A(x) * \mu_B(y)/(x, y)$$

where * is an operator for intersection (Mamdani & Assilian, 1975). On the other hand, if $A \rightarrow B$ is interpreted as A "entails" B, then it can be written as one of two different formulas:

- Material implication:
$$R = A \rightarrow B = \rceil A \cup B .\qquad (2.24)$$

- Propositional Calculus:
$$R = A \rightarrow B = \rceil A \cup (A \cap B) .\qquad (2.25)$$

Although these two formulas are different in appearance, they both reduce to the familiar identity $A \rightarrow B \equiv \rceil A \cup B$ when A and B are propositions in the sense of two-valued logic.

Fuzzy reasoning, also known as approximate reasoning, is an inference procedure that derives conclusions from a set of fuzzy if-then rules and known facts. The basic rule of inference in traditional two-valued logic is "modus ponens", according to which we can infer the truth of a proposition B from the truth of A and the implication $A \rightarrow B$. This concept is illustrated as follows:

premise 1 (fact):	x is A,
premise 2 (rule):	if x is A then y is B,
consequence (conclusion):	y is B.

However, in much of human reasoning, modus ponens is employed in an approximate manner. This is written as

premise 1 (fact):	x is A'
premise 2 (rule):	if x is A then y is B,
consequence (conclusion):	y is B'

where A' is close to A and B' is close to B. When A, B, A' and B' are fuzzy sets of appropriate universes, the foregoing inference procedure is called "approximate reasoning" or "fuzzy reasoning"; it is also called "generalized modus ponens" (GMP for short), since it has modus ponens as a special case.

Definition 2.22. *Fuzzy reasoning*
Let A, A', and B be fuzzy sets of X, X, and Y respectively. Assume that the fuzzy implication $A \rightarrow B$ is expressed as a fuzzy relation R on $X \times Y$. Then the fuzzy set B induced by "x is A" and the fuzzy rule "if x is A then y is B" is defined by

$$\mu_{B'}(y) = \max_x \min[\mu_{A'}(x), \mu_R(x, y)]$$
$$= V_x \left[\mu_{A'}(x) \wedge \mu_R(x, y)\right] . \tag{2.26}$$

Now we can use the inference procedure of fuzzy reasoning to derive conclusions provided that the fuzzy implication $A \rightarrow B$ is defined as an appropriate binary fuzzy relation.

Single Rule with Single Antecedent

This is the simplest case, and the formula is available in (2.26). A further simplification of the equation yields

$$\mu_{B'}(y) = [V_x(\mu_{A'}(x) \wedge \mu_A(x))] \wedge \mu_B(y)$$
$$= \omega \wedge \mu_B(y)$$

In other words, first we find the degree of match ω as the maximum of $\mu_{A'}(x) \wedge \mu_A(x)$; then the MF of the resulting B' is equal to the MF of B clipped by ω. Intuitively, ω represents a measure of degree of belief for the antecedent part of a rule; this measure gets propagated by the if-then rules and the resulting degree of belief or MF for the consequent part should be no greater than ω.

Multiple Rules with Multiple Antecedents

The process of fuzzy reasoning or approximate reasoning for the general case can be divided into four steps:

(1) *Degrees of compatibility*: Compare the known facts with the antecedents of fuzzy rules to find the degrees of compatibility with respect to each antecedent MF.
(2) *Firing strength*: Combine degrees of compatibility with respect to antecedent MFs in a rule using fuzzy AND or OR operators to form a firing strength that indicates the degree to which the antecedent part of the rule is satisfied.
(3) *Qualified (induced) consequent MFs*: Apply the firing strength to the consequent MF of a rule to generate a qualified consequent MF.
(4) *Overall output MF*: Aggregate all the qualified consequent MFs to obtain an overall output MF.

2.3 Fuzzy Inference Systems

In this section we describe the three types of fuzzy inference systems that have been widely used in the applications. The differences between these three fuzzy inference systems lie in the consequents of their fuzzy rules, and thus their aggregation and defuzzification procedures differ accordingly.

The "Mamdani fuzzy inference system" (Mamdani & Assilian, 1975) was proposed as the first attempt to control a steam engine and boiler combination by a set of linguistic control rules obtained from experienced human operators. Figure 2.4 is an illustration of how a two-rule Mamdani fuzzy inference system derives the overall output z when subjected to two numeric inputs x and y.

In Mamdani's application, two fuzzy inference systems were used as two controllers to generate the heat input to the boiler and throttle opening of the engine cylinder, respectively, to regulate the steam pressure in the boiler and the speed of the engine. Since the engine and boiler take only numeric values as inputs, a defuzzifier was used to convert a fuzzy set to a numeric value.

Defuzzification

Defuzzification refers to the way a numeric value is extracted from a fuzzy set as a representative value. In general, there are five methods for defuzzifying a fuzzy set A of a universe of discourse Z, as shown in Fig. 2.5 (Here the fuzzy set A is usually represented by an aggregated output MF, such as C' in Fig. 2.4). A brief explanation of each defuzzification strategy follows.

- Centroid of area z_{COA}:

$$z_{COA} = \frac{\int_z \mu_A(z) z \, dz}{\int_z \mu_A(z) \, dz} \tag{2.27}$$

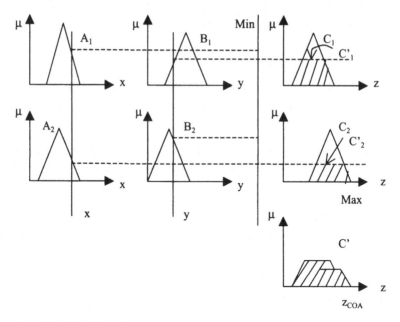

Fig. 2.4. The Mamdani fuzzy inference system using the min and max operators

where $\mu_A(z)$ is the aggregated output MF. This is the most widely adopted defuzzification strategy, which is reminiscent of the calculation of expected values of probability distributions.

- Bisector of area z_{BOA} : z_{BOA} satisfies

$$
\int_\alpha^{z_{BOA}} \mu_A(z)dz = \int_z^\beta \text{BOA } \mu_A(z)dz \;, \tag{2.28}
$$

where $\alpha = \min\{z|z \in Z\}$ and $\beta = \max\{z|z \in Z\}$.

- Mean of maximum z_{MOM} : z_{MOM} is the average of the maximizing z at which the MF reach a maximum μ^*. Mathematically,

$$
z_{MOM} = \frac{\int_{z'} zdz}{\int_{z'} dz} \;, \tag{2.29}
$$

where $z' = \{z|\mu_A(z) = \mu^*\}$. In particular, if $\mu_A(z)$ has a single maximum at $z = z^*$, then $z_{MOM} = z^*$. Moreover, if $\mu_A(z)$ reaches its maximum whenever $z \in [z_{\text{left}}, z_{\text{right}}]$ then $z_{MOM} = (z_{\text{left}} + z_{\text{right}})/2$.

- Smallest of maximum z_{SOM} : z_{SOM} is the minimum (in terms of magnitude) of the maximizing z.
- Largest of maximum z_{LOM} : z_{LOM} is the maximum (in terms of magnitude) of the maximizing z. Because of their obvious bias, z_{SOM} and z_{LOM} are not used as often as the other three defuzzification methods.

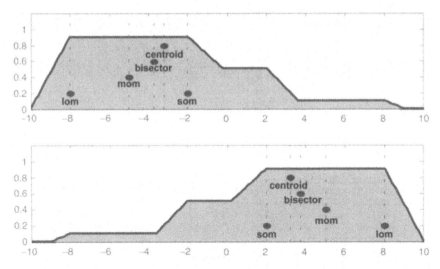

Fig. 2.5. Various defuzzification methods for obtaining a numeric output

The calculation needed to carry out any of these five defuzzification operations is time-consuming unless special hardware support is available. Furthermore, these defuzzification operations are not easily subject to rigorous mathematical analysis, so most of the studies are based on experimental results. This leads to the propositions of other types of fuzzy inference systems that do not need defuzzification at all; two of them will be described in the following. Other more flexible defuzzification methods can be found in several more recent papers (Yager & Filev, 1993), (Runkler & Glesner, 1994).

We will give a simple example to illustrate the use of the Mamdani fuzzy inference system. We will consider the case of determining the quality of a image produce by a Television as a result of controlling the electrical tuning process based on the input variables: voltage and current. Automating the electrical tuning process during the manufacturing of televisions, results in increased productivity and reduction of production costs, as well as increasing the quality of the imaging system of the television. The fuzzy model will consist of a set of rules relating these variables, which represent expert knowledge in the electrical tuning process of televisions. In Fig. 2.6 we show the architecture of the fuzzy system relating the input variables (voltage, current and time) with the output variable (quality of the image), which was implemented by using the MATLAB Fuzzy Logic Toolbox. We show in Fig. 2.7 the fuzzy rule base, which was implemented by using the "rule editor" of the same toolbox. In Fig. 2.8 we can appreciate the membership functions for the image-quality variable. We show in Fig. 2.9 the membership functions for the voltage variable. We also show in Fig. 2.10 the use of the "rule viewer" of MATLAB to calculate specific values. Finally, in Fig. 2.11 we show the non-linear surface for the Mamdani model.

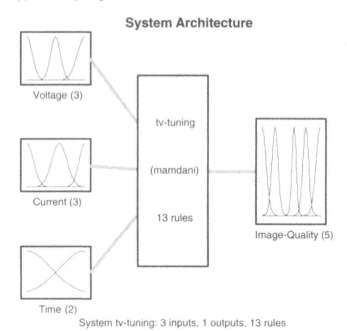

System Architecture

Voltage (3)

Current (3)

Time (2)

tv-tuning

(mamdani)

13 rules

Image-Quality (5)

System tv-tuning: 3 inputs, 1 outputs, 13 rules

Fig. 2.6. Architecture of the fuzzy system for quality evaluation

Sugeno Fuzzy Models

The "Sugeno fuzzy model" (also known as the "TSK fuzzy model") was proposed by Takagi, Sugeno and Kang in an effort to develop a systematic approach to generating fuzzy rules from a given input-output data set (Takagi & Sugeno, 1985), (Sugeno & Kang, 1988). A typical fuzzy rule in a Sugeno fuzzy model has the form:

$$\textbf{if } x \textbf{ is } A \textbf{ and } y \textbf{ is } B \textbf{ then } z = f(x,y)$$

where A and B are fuzzy sets in the antecedent, while $z = f(x,y)$ is a traditional function in the consequent. Usually $f(x,y)$ is a polynomial in the input variables x and y, but it can be any function as long as it can appropriately describe the output of the model within the fuzzy region specified by the antecedent of the rule. When $f(x,y)$ is a first-order polynomial, the resulting fuzzy inference system is called a "first-order Sugeno fuzzy model". When f is constant, we then have a "zero-order Sugeno fuzzy model", which can be viewed either as a special case of the Mamdani inference system, in which each rule's consequent is specified by a fuzzy singleton, or a special case of the Tsukamoto fuzzy model (to be introduced next), in which each rule's consequent is specified by a MF of a step function center at the constant.

Figure 2.12 shows the fuzzy reasoning procedure for a first-order Sugeno model. Since each rule has a numeric output, the overall output is obtained via

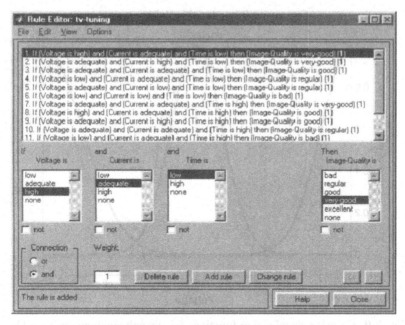

Fig. 2.7. Fuzzy rule base for quality evaluation

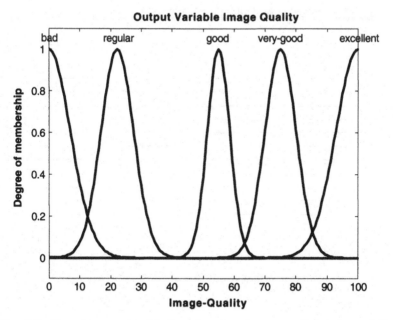

Fig. 2.8. Gaussian membership functions for the output linguistic variable

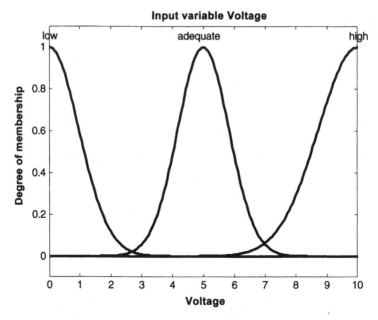

Fig. 2.9. Gaussian membership functions for the voltage linguistic variable

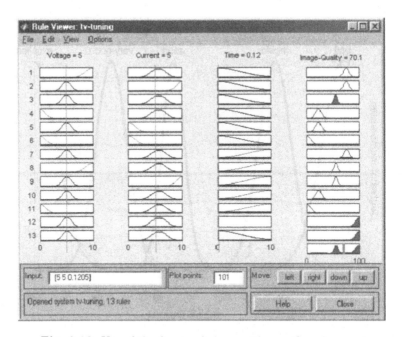

Fig. 2.10. Use of the fuzzy rule base with specific values

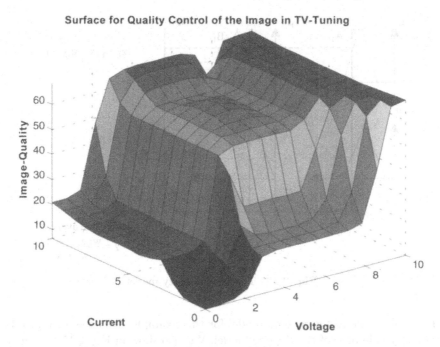

Surface for Quality Control of the Image in TV-Tuning

Fig. 2.11. Non-linear surface of the Mamdani fuzzy model

"weighted average", thus avoiding the time-consuming process of defuzzification required in a Mamdani model. In practice, the weighted average operator is sometimes replaced with the "weighted sum" operator (that is, $w_1 z_1 + w_2 z_2$ in Fig. 2.12) to reduce computation further specially, in the training of a fuzzy inference system. However, this simplification could lead to the loss of MF linguistic meanings unless the sum of firing strengths (that is, $\sum w_i$) is close to unity.

Unlike the Mamdani fuzzy model, the Sugeno fuzzy model cannot follow the compositional rule of inference strictly in its fuzzy reasoning mechanism. This poses some difficulties when the inputs to a Sugeno fuzzy model are fuzzy. Specifically, we can still employ the matching of fuzzy sets to find the firing strength of each rule. However, the resulting overall output via either weighted average or weighted sum is always crisp; this is counterintuitive since a fuzzy model should be able to propagate the fuzziness from inputs to outputs in an appropriate manner. Without the use of the time-consuming defuzzification procedure, the Sugeno fuzzy model is by far the most popular candidate for sample-data-based modeling.

We will give a simple example to illustrate the use of the Sugeno fuzzy inference system. We will consider again the television example, i.e., determining the quality of the images produced by the television depending on the voltage and current of the electrical tuning process. In Fig. 2.13 we show

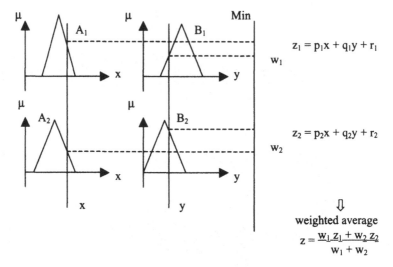

$$z_1 = p_1 x + q_1 y + r_1$$
w_1

$$z_2 = p_2 x + q_2 y + r_2$$
w_2

⇓

weighted average

$$z = \frac{w_1 z_1 + w_2 z_2}{w_1 + w_2}$$

Fig. 2.12. The Sugeno fuzzy model

the architecture of the Sugeno model for this example. We show in Fig. 2.14 the fuzzy rule base of the Sugeno model. We also show in Fig. 2.15 the membership functions for the current input variable. In Fig. 2.16 we show the non-linear surface of the Sugeno model.

Finally, we show in Fig. 2.17 the use of the "rule viewer" of the Fuzzy Logic Toolbox of MATLAB. The rule viewer is used when we want to evaluate the output of a fuzzy system using specific values for the input variables. In Fig. 2.17, for example, we give a voltage of 5 volts, a current intensity of 5 Amperes, and a time of production of 5 seconds, and obtain as a result a quality of 92.2%, which is excellent. Of course, this is only an illustrative example of the potential use of fuzzy logic in this type of applications.

Tsukamoto Fuzzy Models

In the "Tsukamoto fuzzy models" (Tsukamoto, 1979), the consequent of each fuzzy if-then rule is represented by a fuzzy set with a monotonical MF, as shown in Fig. 2.18. As a result, the inferred output of each rule is defined as a numeric value induced by the rule firing strength. The overall output is taken as the weighted average of each rule's output. Figure 2.18 illustrates the reasoning procedure for a two-input two-rule system.

Since each rule infers a numeric output, the Tsukamoto fuzzy model aggregates each rule's output by the method of weighted average and thus avoids the time-consuming process of defuzzification. However, the Tsukamoto fuzzy model is not used often since it is not as transparent as either the Mamdani or Sugeno fuzzy models. Since the reasoning method of the Tsukamoto fuzzy

System Architecture for Sugeno Type

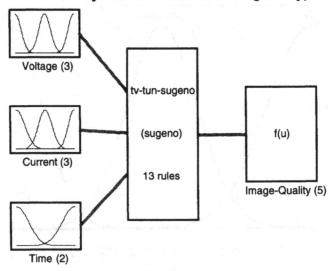

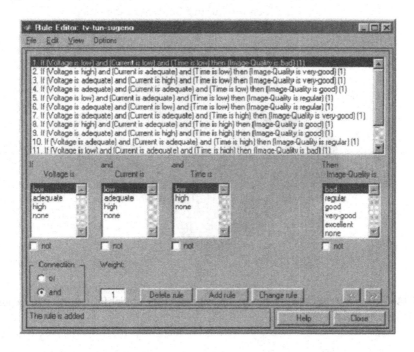

System tv-tun-sugeno: 3 inputs, 1 outputs, 13 rules

Fig. 2.13. Architecture of the Sugeno fuzzy model for quality evaluation

Fig. 2.14. Fuzzy rule base for quality evaluation using the "rule editor"

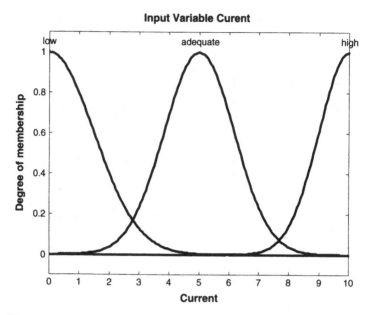

Fig. 2.15. Membership functions for the current linguistic variable

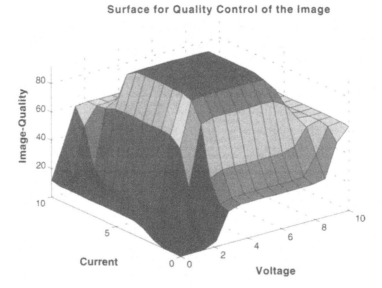

Fig. 2.16. Non-linear surface for the Sugeno fuzzy model for quality evaluation

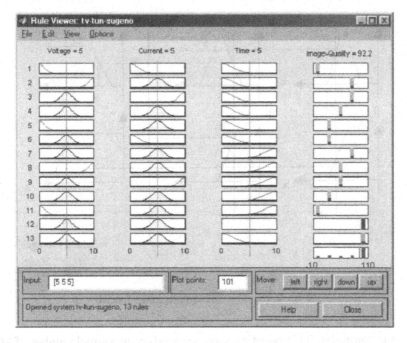

Fig. 2.17. Application of the rule viewer of MATLAB with specific values

model does not follow strictly the compositional rule of inference, the output is always crisp even when the inputs are fuzzy.

There are certain common issues concerning all the three fuzzy inference systems introduced previously, such as how to partition an input space and how to construct a fuzzy inference system for a particular application. We will examine these issues in more detail in the following lines.

Input Space Partitioning

Now it should be clear that the main idea of fuzzy inference systems resembles that of "divide and conquer" – the antecedent of a fuzzy rule defines a local fuzzy region, while the consequent describes the behavior within the region via various constituents. The consequent constituent can be a consequent MF (Mamdani and Tsukamoto fuzzy models), a constant value (zero-order Sugeno model), a linear equation (first-order Sugeno model) or a non-linear equation (higher order Sugeno models). Different consequent constituents result in different fuzzy inference systems, but their antecedents are always the same. Therefore, the following discussion of methods of partitioning input spaces to form the antecedents of fuzzy rules is applicable to all three types of fuzzy inference systems.

- Grid partition: This partition method is often chosen in designing a fuzzy controller, which usually involves only several state variables as the inputs

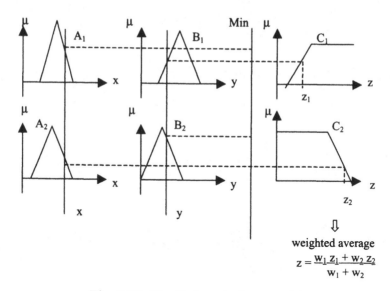

Fig. 2.18. The Tsukamoto fuzzy model

to the controller. This partition strategy needs only a small number of MFs for each input. However, it encounters problems when we have a moderately large number of inputs. For instance, a fuzzy model with 12 inputs and 2 MFs on each input would result in $2^{12} = 4096$ fuzzy if-then rules, which is prohibitively large. This problem, usually referred to as the "curse of dimensionality", can be alleviated by other partition strategies.

- Tree partition: In this method each region can be uniquely specified along a corresponding decision tree. The tree partition relieves the problem of an exponential increase in the number of rules. However, more MFs for each input are needed to define these fuzzy regions, and these MFs do not usually bear clear linguistic meanings. In other words, orthogonality holds roughly in $X \times Y$, but not in either X or Y alone.

- Scatter partition: By covering a subset of the whole input space that characterizes a region of possible occurrence of the input vectors, the scatter partition can also limit the number of rules to a reasonable amount. However, the scatter partition is usually dictated by desired input-output data pairs and thus, in general, orthogonality does not hold in X, Y or $X \times Y$. This makes it hard to estimate the overall mapping directly from the consequent of each rule's output.

2.4 Fuzzy Modeling

In general, we design a fuzzy inference system based on the past known behavior of a target system. The fuzzy system is then expected to be able to reproduce the behavior of the target system. For example, if the target system is a human operator in charge of a electrochemical reaction process, then the fuzzy inference system becomes a fuzzy logic controller that can regulate and control the process.

Let us now consider how we might construct a fuzzy inference system for a specific application. Generally speaking, the standard method for constructing a fuzzy inference system, a process usually called "fuzzy modeling", has the following features:

- The rule structure of a fuzzy inference system makes it easy to incorporate human expertise about the target system directly into the modeling process. Namely, fuzzy modeling takes advantage of "domain knowledge" that might not be easily or directly employed in other modeling approaches.
- When the input-output data of a target system is available, conventional system identification techniques can be used for fuzzy modeling. In other words, the use of "numerical data" also plays an important role in "fuzzy modeling", just as in other mathematical modeling methods.

Conceptually, fuzzy modeling can be pursued in two stages, which are not totally disjoint. The first stage is the identification of the "surface structure", which includes the following tasks:

1. Select relevant input and output variables.
2. Choose a specific type of fuzzy inference system.
3. Determine the number of linguistic terms associated with each input and output variables.
4. Design a collection of fuzzy if-then rules.

Note that to accomplish the preceding tasks, we rely on our own knowledge (common sense, simple physical laws, an so on) of the target system, information provided by human experts who are familiar with the target system, or simply trial and error.

After the first stage of fuzzy modeling, we obtain a rule base that can more or less describe the behavior of the target system by means of linguistic terms. The meaning of these linguistic terms is determined in the second stage, the identification of "deep structure", which determines the MFs of each linguistic term (and the coefficients of each rule's output in the case that a Sugeno model is used). Specifically, the identification of deep structure includes the following tasks:

1. Choose an appropriate family of parameterized MFs.
2. Interview human experts familiar with the target systems to determine the parameters of the MFs used in the rule base.
3. Refine the parameters of the MFs using regression and optimization techniques.

Task 1 and 2 assume the availability of human experts, while task 3 assumes the availability of a desired input-output data set. When a fuzzy inference system is used as a controller for a given plant, then the objective in task 3 should be changed to that of searching for parameters that will generate the best performance of the plant.

2.5 Summary

In this chapter, we have presented the main ideas underlying type-1 fuzzy logic and we have only started to point out the many possible applications of this powerful computational theory. We have discussed in some detail fuzzy set theory, fuzzy reasoning and fuzzy inference systems. At the end, we also gave some remarks about fuzzy modeling. In the following chapters, we will show how fuzzy logic techniques (in some cases, in conjunction with other methodologies) can be applied to solve real world complex problems. This chapter will serve as a basis for the new hybrid intelligent methods, for pattern recognition that will be described later in this book. Fuzzy Logic will also play an important role in the new neuro-fuzzy and fuzzy-genetic methodologies for pattern recognition that are presented later in this book. Applications of fuzzy logic to build intelligent systems for pattern recognition will be described in Chaps. 9, 10, 11, and 12.

3

Intuitionistic and Type-2 Fuzzy Logic

We describe in this chapter two new areas in fuzzy logic, type-2 fuzzy logic systems and intuitionistic fuzzy logic. Basically, a type-2 fuzzy set is a set in which we also have uncertainty about the membership function. Of course, type-2 fuzzy systems consist of fuzzy if-then rules, which contain type-2 fuzzy sets. We can say that type-2 fuzzy logic is a generalization of conventional fuzzy logic (type-1) in the sense that uncertainty is not only limited to the linguistic variables but also is present in the definition of the membership functions. On the other hand, intuitionistic fuzzy sets can also be considered an extension of type-1 fuzzy sets in the sense that intuitionistic fuzzy sets not only use the membership function, but also a non-membership function to represent the uncertainty of belonging to a fuzzy set.

Fuzzy Logic Systems are comprised of rules. Quite often, the knowledge that is used to build these rules is uncertain. Such uncertainty leads to rules whose antecedents or consequents are uncertain, which translates into uncertain antecedent or consequent membership functions (Karnik & Mendel 1998). Type-1 fuzzy systems (like the ones seen in the previous chapter), whose membership functions are type-1 fuzzy sets, are unable to directly handle such uncertainties. We describe in this chapter, type-2 fuzzy systems, in which the antecedent or consequent membership functions are type-2 fuzzy sets. Such sets are fuzzy sets whose membership grades themselves are type-1 fuzzy sets; they are very useful in circumstances where it is difficult to determine an exact membership function for a fuzzy set.

The original fuzzy logic, founded by Lotfi Zadeh, has been around for more than 30 years, and yet it is unable to handle uncertainties (Mendel, 2001). That the original fuzzy logic (type-1 fuzzy logic) cannot do this sounds paradoxical because the word "fuzzy" has the connotation of uncertainty. The expanded fuzzy logic (type-2 fuzzy logic) is able to handle uncertainties because it can model and minimize their effects.

In what follows, we shall first introduce the basic concepts of type-2 fuzzy sets, and type-2 fuzzy reasoning. Then we will introduce and compare the different types of fuzzy inference systems that have been employed in various

Patricia Melin and Oscar Castillo: *Hybrid Intelligent Systems for Pattern Recognition Using Soft Computing*, StudFuzz **172**, 33–53 (2005)
www.springerlink.com   © Springer-Verlag Berlin Heidelberg 2005

applications. We will also consider briefly type-2 fuzzy logic systems and the comparison to type-1 fuzzy systems. Then we will describe the concept of an intuitionistic fuzzy set and its applications. We will also describe intuitionistic fuzzy inference systems. Finally, we will address briefly the features and problems of fuzzy modeling with intuitionistic and type-2 fuzzy logic, which is concerned with the construction of fuzzy inference systems for modeling a given target system.

3.1 Type-2 Fuzzy Sets

The concept of a type-2 fuzzy set, was introduced by Zadeh (1975) as an extension of the concept of an ordinary fuzzy set (henceforth called a "type-1 fuzzy set"). A type-2 fuzzy set is characterized by a fuzzy membership function, i.e., the membership grade for each element of this set is a fuzzy set in [0, 1], unlike a type-1 set where the membership grade is a crisp number in [0, 1]. Such sets can be used in situations where there is uncertainty about the membership grades themselves, e.g., an uncertainty in the shape of the membership function or in some of its parameters. Consider the transition from ordinary sets to fuzzy sets. When we cannot determine the membership of an element in a set as 0 or 1, we use fuzzy sets of type-1. Similarly, when the situation is so fuzzy that we have trouble determining the membership grade even as a crisp number in [0, 1], we use fuzzy sets of type-2.

This does not mean that we need to have extremely fuzzy situations to use type-2 fuzzy sets. There are many real-world problems where we cannot determine the exact form of the membership functions, e.g., in time series prediction because of noise in the data. Another way of viewing this is to consider type-1 fuzzy sets as a first order approximation to the uncertainty in the real-world. Then type-2 fuzzy sets can be considered as a second order approximation. Of course, it is possible to consider fuzzy sets of higher types but the complexity of the fuzzy system increases very rapidly. For this reason, we will only consider very briefly type-2 fuzzy sets. Lets consider some simple examples of type-2 fuzzy sets.

Example 3.1. Consider the case of a fuzzy set characterized by a Gaussian membership function with mean m and a standard deviation that can take values in $[\sigma_1, \sigma_2]$, i.e.,

$$\mu(x) = \exp\{-1/2[(x-m)/\sigma]^2\}; \quad \sigma \in [\sigma_1, \sigma_2] \tag{3.1}$$

Corresponding to each value of σ, we will get a different membership curve (see Fig. 3.1). So, the membership grade of any particular x (except $x = m$) can take any of a number of possible values depending upon the value of σ, i.e., the membership grade is not a crisp number, it is a fuzzy set. Figure 3.1 shows the domain of the fuzzy set associated with $x = 0.7$; however, the membership function associated with this fuzzy set is not shown in the figure.

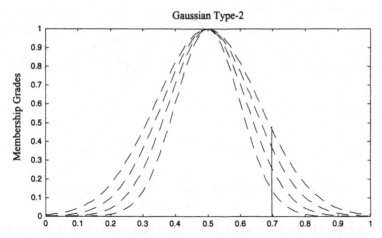

Fig. 3.1. A type-2 fuzzy set representing a type-1 fuzzy set with uncertain standard deviation

Example 3.2. Consider the case of a fuzzy set with a Gaussian membership function having a fixed standard deviation σ, but an uncertain mean, taking values in $[m_1, m_2]$, i.e.,

$$\mu(x) = \exp\left\{-1/2[(x-m)/\sigma]^2\right\}; \quad m \in [m_1, m_2] \tag{3.2}$$

Again, $\mu(x)$ is a fuzzy set. Figure 3.2 shows an example of such a set.

Example 3.3. Consider a type-1 fuzzy set characterized by a Gaussian membership function (mean M and standard deviation σ_x), which gives one crisp membership $m(x)$ for each input $x \in X$, where

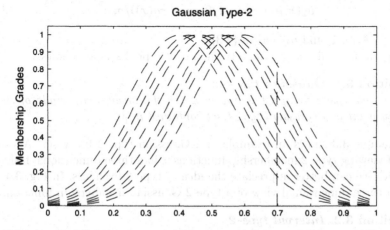

Fig. 3.2. A type-2 fuzzy set representing a type-1 fuzzy set with uncertain mean. The mean is uncertain in the interval [0.4, 0.6]

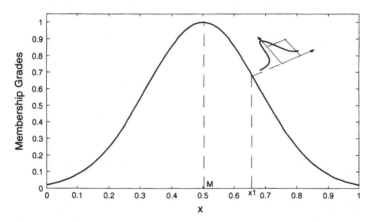

Fig. 3.3. A type-2 fuzzy set in which the membership grade of every domain point is a Gaussian type-1 set

$$m(x) = \exp\left\{-1/2[(x - M)/\sigma_x]^2\right\} \qquad (3.3)$$

This is shown in Fig. 3.3. Now, imagine that this membership of x is a fuzzy set. Let us call the domain elements of this set "primary memberships" of x (denoted by μ_1) and membership grades of these primary memberships "secondary memberships" of x [denoted by $\mu_2(x, \mu_1)$]. So, for a fixed x, we get a type-1 fuzzy set whose domain elements are primary memberships of x and whose corresponding membership grades are secondary memberships of x. If we assume that the secondary memberships follow a Gaussian with mean $m(x)$ and standard deviation σ_m, as in Fig. 3.3, we can describe the secondary membership function for each x as

$$\mu_2(x, \mu_1) = e - 1/2\left[(\mu_1 - m(x))/\sigma_m\right]^2 \qquad (3.4)$$

where $\mu_1 \in [0, 1]$ and m is as in (3.3).

We can formally define these two kinds of type-2 sets as follows.

Definition 3.1. *Gaussian type-2*
A Gaussian type-2 fuzzy set is one in which the membership grade of every domain point is a Gaussian type-1 set contained in [0, 1].

Example 3.3 shows an example of a Gaussian type-2 fuzzy set. Another way of viewing type-2 membership functions is in a three-dimensional fashion, in which we can better appreciate the idea of type-2 fuzziness. In Fig. 3.4 we have a three-dimensional view of a type-2 Gaussian membership function.

Definition 3.2. *Interval type-2*
An interval type-2 fuzzy set is one in which the membership grade of every domain point is a crisp set whose domain is some interval contained in [0, 1].

Type-2 fuzzy set in 3D

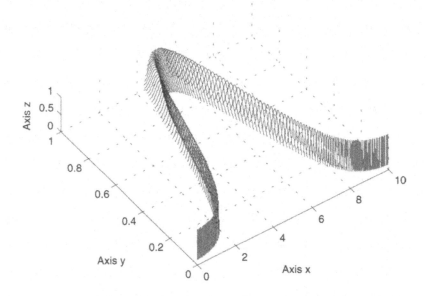

Fig. 3.4. Three-dimensional view of a type-2 membership function

Example 3.1 shows an example of an interval type-2 fuzzy set.

We will give some useful definitions on type-2 fuzzy sets in the following lines.

Definition 3.3. *Footprint of uncertainty*
Uncertainty in the primary memberships of a type-2 fuzzy set, Ã, consists of a bounded region that we call the "footprint of uncertainty" (FOU). Mathematically, it is the union of all primary membership functions (Mendel, 2001).

We show as an illustration in Fig. 3.5 the footprint of uncertainty for a type-2 Gaussian membership function. This footprint of uncertainty can be obtained by projecting in two dimensions the three-dimensional view of the type-2 Gaussian membership function.

Definition 3.4. *Upper and lower membership functions*
An "upper membership function" and a "lower membership functions" are two type-1 membership functions that are bounds for the FOU of a type-2 fuzzy set Ã. The upper membership function is associated with the upper bound of FOU(Ã). The lower membership function is associated with the lower bound of FOU(Ã).

We illustrate the concept of upper and lower membership functions as well as the footprint of uncertainty in the following example.

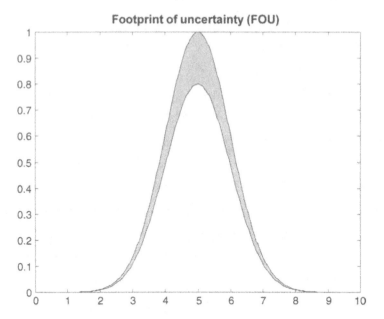

Fig. 3.5. Footprint of uncertainty of a sample type-2 Gaussian membership function

Example 3.4. Gaussian primary MF with uncertain standard deviation
For the Gaussian primary membership function with uncertain standard deviation (Fig. 3.1), the upper membership function is

$$\text{upper}(\text{FOU}(\tilde{A})) = N(m, \sigma_2; x) \tag{3.5}$$

And the lower membership function is

$$\text{lower}(\text{FOU}(\tilde{A})) = N(m, \sigma_1; x) . \tag{3.6}$$

We will describe the operations and properties of type-2 fuzzy sets in the following section.

3.2 Operations of Type-2 Fuzzy Sets

In this section we describe the set theoretic operations of type-2 fuzzy sets. We are interested in the case of type-2 fuzzy sets, $\tilde{A}_i$ $(i = 1, \ldots, r)$, whose secondary membership functions are type-1 fuzzy sets. To compute the union, intersection, and complement of type-2 fuzzy sets, we need to extend the binary operations of minimum (or product) and maximum, and the unary operation of negation, from crisp numbers to type-1 fuzzy sets, because at each x, $\mu_{\tilde{A}i}(x, u)$ is a function (unlike the type-1 case, where $\mu_{\tilde{A}i}(x)$ is a crisp number). The tool for computing the union, intersection, and complement of type-2 fuzzy sets is Zadeh's extension principle (Zadeh, 1975).

Consider two type-2 fuzzy sets $\tilde{A}_1$ and $\tilde{A}_2$, i.e.,

$$\tilde{A}_1 = \int_x \mu_{\tilde{A}1}(x)/x \tag{3.7}$$

and

$$\tilde{A}_2 = \int_x \mu_{\tilde{A}2}(x)/x \tag{3.8}$$

In this section, we focus our attention on set theoretic operations for such general type-2 fuzzy sets.

Definition 3.5. *Union of type-2 fuzzy sets*
The union of $\tilde{A}_1$ and $\tilde{A}_2$ is another type-2 fuzzy set, just as the union of type-1 fuzzy sets A_1 and A_2 is another type-1 fuzzy set. More formally, we have the following expression

$$\tilde{A}_1 \cup \tilde{A}_2 = \int_{x \in X} \mu_{\tilde{A}1 \cup \tilde{A}2}(x)/x \tag{3.9}$$

We can explain (3.9) by the "join" operation (Mendel, 2001). Basically, the join between two secondary membership functions must be performed between every possible pair of primary memberships. If more than one combination of pairs gives the same point, then in the join we keep the one with maximum membership grade. We will consider a simple example to illustrate the union operation. In Fig. 3.6 we plot two type-2 Gaussian membership functions, and the union is shown in Fig. 3.7.

Definition 3.6. *Intersection of type-2 fuzzy sets*
The intersection of $\tilde{A}_1$ and $\tilde{A}_2$ is another type-2 fuzzy set, just as the intersection of type-1 fuzzy sets A_1 and A_2 is another type-1 fuzzy set. More formally, we have the following expression

$$\tilde{A}_1 \cap \tilde{A}_2 = \int_{x \in X} \mu_{\tilde{A}1 \cap \tilde{A}2}(x)/x \tag{3.10}$$

We illustrate the intersection of two type-2 Gaussian membership functions in Fig. 3.8.

We can explain (3.10) by the "meet" operation (Mendel, 2001). Basically, the meet between two secondary membership functions must be performed between every possible pair of primary memberships. If more than one combination of pairs gives the same point, then in the meet we keep the one with maximum membership grade.

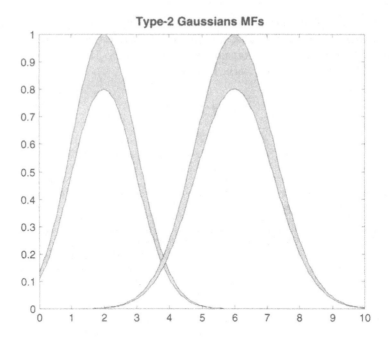

Fig. 3.6. Two sample type-2 Gaussian membership functions

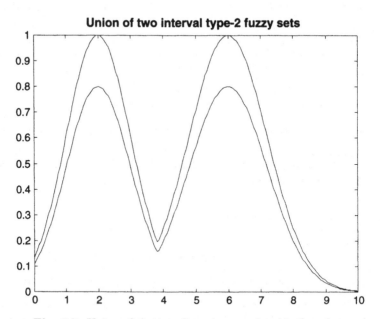

Fig. 3.7. Union of the two Gaussian membership functions

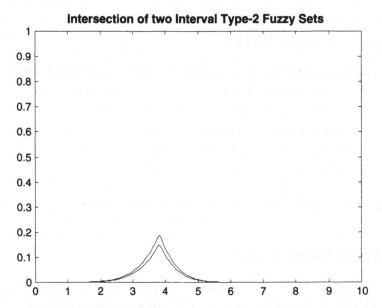

Fig. 3.8. Intersection of two type-2 Gaussian membership functions

Definition 3.7. *Complement of a type-2 fuzzy set*

The complement of set $\tilde{A}$ is another type-2 fuzzy set, just as the complement of type-1 fuzzy set A is another type-1 fuzzy set. More formally we have

$$\tilde{A}' = \int_x \mu_{\tilde{A}'1}(x)/x \qquad (3.11)$$

where the prime denotes complement in the above equation. In this equation $\mu_{\tilde{A}'1}$ is a secondary membership function, i.e., at each value of $x\,\mu_{\tilde{A}'1}$ is a function (unlike the type-1 case where, at each value of $x, \mu_{\tilde{A}'1}$ is a point value).

Example 3.5. Type-2 fuzzy set operations
In this example we illustrate the union, intersection and complement operations for two type-2 fuzzy sets $\tilde{A}_1$ and $\tilde{A}_2$, and for a particular element x for which the secondary membership functions in these two sets are $\mu_{\tilde{A}1}(x) = 0.5/0.1+0.8/0.2$ and $\mu_{\tilde{A}2}(x) = 0.4/0.5 + 0.9/0.9$. Using in the operations the minimum t-norm and the maximum t-conorm, we have the following results:

$$
\begin{aligned}
\mu_{\tilde{A}1\cup\tilde{A}2}(x) &= \mu_{\tilde{A}1}(x)\cup\mu_{\tilde{A}2}(x) = (0.5/0.1 + 0.8/0.2)\cup(0.4/0.5 + 0.9/0.9) \\
&= (0.5\wedge 0.4)/(0.1\vee 0.5) + (0.5\wedge 0.9)/(0.1\vee 0.9) \\
&\quad +(0.8\wedge 0.4)/(0.2\vee 0.5) + (0.8\wedge 0.9)/(0.2\vee 0.9) \\
&= 0.4/0.5 + 0.5/0.9 + 0.4/0.5 + 0.8/0.9
\end{aligned}
$$

$$= \max\{0.4, 0.4\}/0.5 + \max\{0.5, 0.8\}/0.9$$
$$= 0.4/0.5 + 0.8/0.9$$

$$\mu_{\tilde{A}1 \cap \tilde{A}2}(x) = \mu_{\tilde{A}1}(x) \cap \mu_{\tilde{A}2}(x) = (0.5/0.1 + 0.8/0.2) \cap (0.4/0.5 + 0.9/0.9)$$
$$= (0.5 \wedge 0.4)/(0.1 \wedge 0.5) + (0.5 \wedge 0.9)/(0.1 \wedge 0.9)$$
$$+ (0.8 \wedge 0.4)/(0.2 \wedge 0.5) + (0.8 \wedge 0.9)/(0.2 \wedge 0.9)$$
$$= 0.4/0.1 + 0.5/0.1 + 0.4/0.2 + 0.8/0.2$$
$$= \max\{0.4, 0.5\}/0.1 + \max\{0.4, 0.8\}/0.2$$
$$= 0.5/0.1 + 0.8/0.2$$

$$\mu_{\tilde{A}'1}(x) = 0.5/(1 - 0.1) + 0.8/(1 - 0.2) = 0.5/0.9 + 0.8/0.8 .$$

3.3 Type-2 Fuzzy Systems

The basics of fuzzy logic do not change from type-1 to type-2 fuzzy sets, and in general, will not change for any type-n (Karnik & Mendel 1998). A higher-type number just indicates a higher "degree of fuzziness". Since a higher type changes the nature of the membership functions, the operations that depend on the membership functions change; however, the basic principles of fuzzy logic are independent of the nature of membership functions and hence, do not change. Rules of inference like Generalized Modus Ponens or Generalized Modus Tollens continue to apply.

The structure of the type-2 fuzzy rules is the same as for the type-1 case because the distinction between type-2 and type-1 is associated with the nature of the membership functions. Hence, the only difference is that now some or all the sets involved in the rules are of type-2. In a type-1 fuzzy system, where the output sets are type-1 fuzzy sets, we perform defuzzification in order to get a number, which is in some sense a crisp (type-0) representative of the combined output sets. In the type-2 case, the output sets are type-2; so we have to use extended versions of type-1 defuzzification methods. Since type-1 defuzzification gives a crisp number at the output of the fuzzy system, the extended defuzzification operation in the type-2 case gives a type-1 fuzzy set at the output. Since this operation takes us from the type-2 output sets of the fuzzy system to a type-1 set, we can call this operation "type reduction" and call the type-1 fuzzy set so obtained a "type-reduced set". The type-reduced fuzzy set may then be defuzzified to obtain a single crisp number; however, in many applications, the type-reduced set may be more important than a single crisp number.

Type-2 sets can be used to convey the uncertainties in membership functions of type-1 fuzzy sets, due to the dependence of the membership functions on available linguistic and numerical information. Linguistic information (e.g. rules from experts), in general, does not give any information about the shapes of the membership functions. When membership functions are determined or tuned based on numerical data, the uncertainty in the numerical data, e.g., noise, translates into uncertainty in the membership functions. In all such cases, any available information about the linguistic/numerical uncertainty can be incorporated in the type-2 framework. However, even with all of the advantages that fuzzy type-2 systems have, the literature on the applications of type-2 sets is scarce. Some examples are Yager (1980) for decision making, and Wagenknecht & Hartmann (1998) for solving fuzzy relational equations. We think that more applications of type-2 fuzzy systems will come in the near future as the area matures and the theoretical results become more understandable for the general public in the fuzzy arena.

3.3.1 Singleton Type-2 Fuzzy Logic Systems

This section discusses the structure of a singleton type-2 Fuzzy Logic Systems (FLS), which is a system that accounts for uncertainties about the antecedents or consequents in rules, but does not explicitly account for input measurement uncertainties. More complicated (but, more versatile) non-singleton type-2 FLSs, which account for both types of uncertainties, are discussed later.

The basics of fuzzy logic do not change from type-1 to type-2 fuzzy sets, and in general will not change for type-n. A higher type number just indicates a higher degree of fuzziness. Since a higher type changes the nature of the membership functions, the operations that depend on the membership functions change, however, the basic principles of fuzzy logic are independent of the nature of membership functions and hence do not change. Rules of inference, like Generalized Modus Ponens, continue to apply.

A general type-2 FLS is shown in Fig. 3.9. As discussed before a type-2 FLS is very similar to type-1 FLS, the major structural difference being that the defuzzifier block of a type-1 FLS is replaced by the output processing block in type-2 FLS. That block consists of type-reduction followed by defuzzification.

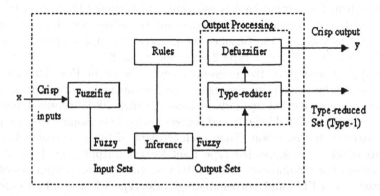

Fig. 3.9. Type-2 Fuzzy Logic System

During our development of a type-2 FLS, we assume that all the antecedent and consequent sets in rules are type-2, however, this need not necessarily be the case in practice. All results remain valid as long as long as just one set is type-2. This means that a FLS is type-2 as long as any one of its antecedent or consequent sets is type-2.

In the type-1 case, we generally have fuzzy if-then rules of the form

$$R^l : \text{ IF } x_1 \text{ is } A_1^l \text{ and } \ldots x_p \text{ is } A_p^l, \text{ THEN } y \text{ is } Y^l \quad l = 1, \ldots, M \quad (3.12)$$

As mentioned earlier, the distinction between type-1 and type-2 is associated with the nature of the membership functions, which is not important when forming the rules. The structure of the rules remains exactly the same in the type-2 case, but now some or all of the sets involved are type-2.

Consider a type-2 FLS having r inputs $x_1 \in X_1, \ldots, x_r \in X_r$ and one output $y \in Y$. As in the type-1 case, we can assume that there are M rules; but, in the type-2 case the lth rule has the form

$$R^l : \text{ IF } x_1 \text{ is } \tilde{A}_1^l \text{ and } \ldots x_p \text{ is } \tilde{A}_p^l, \text{ THEN } y \text{ is } \hat{Y}^l \quad 1 = 1, \ldots, M \quad (3.13)$$

This rule represents a type-2 fuzzy relation between the input space $X_1 \times \ldots \times X_r$, and the output space, Y, of the type-2 fuzzy system.

In a type-1 FLS the inference engine combines rules and gives a mapping from input type-1 fuzzy sets to output type-1 fuzzy sets. Multiple antecedents in rules are combined by the t-norm. The membership grades in the input sets are combined with those in the output sets using composition. Multiple rules may be combined using the t-conorm or during defuzzification by weighted summation. In the type-2 case the inference process is very similar. The inference engine combines rules and gives a mapping from input type-2 fuzzy sets to output type-2 fuzzy sets. To do this one needs to compute unions and intersections of type-2 fuzzy sets, as well as compositions of type-2 relations.

In the type-2 fuzzy system (Fig. 3.9), as in the type-1 fuzzy system, crisp inputs are first fuzzified into fuzzy input sets that then activate the inference block, which in the present case is associated with type-2 fuzzy sets. In this section, we describe singleton fuzzification and the effect of such fuzzification on the inference engine. The "fuzzifier" maps a crisp point $\mathbf{x} = (x_1, \ldots, x_r)^T \in X_1 \times X_2 \ldots \times X_r \equiv \mathbf{X}$ into a type-2 fuzzy set $\tilde{A}_x$ in $\mathbf{X}$.

The type-2 output of the inference engine shown in Fig. 3.9 must be processed next by the output processor, the first operation of which is type-reduction. Type-reduction methods include (Mendel, 2001): centroid, center-of-sums, height, modified height, and center-of-sets. Lets assume that we perform centroid type-reduction. Then each element of the type-reduced set is the centroid of some embedded type-1 set for the output type-2 set of the FLS. Each of these embedded sets can be thought of as an output set of an associated type-1 FLS, and, correspondingly, the type-2 FLS can be viewed of as a collection of many different type-1 FLSs. Each type-1 FLS is embedded in the type-2 FLS; hence, the type-reduced set is a collection of the outputs

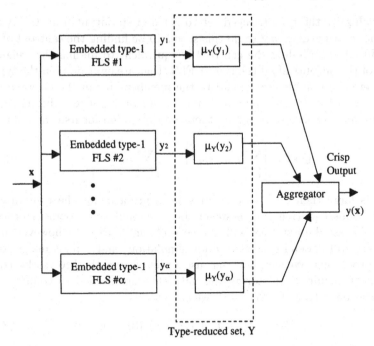

Fig. 3.10. A type-2 FLS viewed as a collection of embedded type-1 FLSs

of all of the embedded type-1 FLSs (see Fig. 3.10). The type-reduced set lets us represent the output of the type-2 FLS as a fuzzy set rather than as a crisp number, which is something that cannot be done with a type-1 fuzzy system.

Referring to Fig. 3.10, when the antecedent and consequent membership functions of the type-2 FLS have continuous domains, the number of embedded sets is uncountable. Figure 3.10 shows a situation in which we have assumed that the membership functions have discrete (or discretized) domains. The memberships in the type-reduced set, $\mu_Y(y_i)$, represent the level of uncertainty associated with each embedded type-1 FLS. A crisp output can be obtained by aggregating the outputs of all embedded type-1 FLSs by, e.g., finding the centroid of the type-reduced set.

If all of the type-2 uncertainties were to disappear, the secondary membership functions for all antecedents and consequents would each collapse to a single point, which shows that the type-2 FLS collapses to a type-1 FLS.

If we think of a type-2 FLS as a "perturbation" of a type-1 FLS, due to uncertainties in their membership functions, then the type-reduced set of the type-2 FLS can be thought of as representing the uncertainty in the crisp output due to the perturbation. Some measure of the spread of the type-reduced set may then be taken to indicate the possible variation in the crisp output due to the perturbation. This is analogous to using confidence intervals in a stochastic-uncertainty situation.

We defuzzify the type-reduced set to get a crisp output from the type-2 FLS. The most natural way to do this seems to be finding the centroid of the type-reduced set. Finding the centroid is equivalent to finding the weighted average of the outputs of all the type-1 FLSs that are embedded in the type-2 FLS, where the weights correspond to the memberships in the type-reduced set (see Fig. 3.10). If the type-reduced set Y for an input $\mathbf{x}$ is discretized or is discrete and consists of α points, then the expression for its centroid is

$$y(\mathbf{x}) = \left[\sum_{k=1}^{\alpha} y_k \mu_y(y_k)\right] \bigg/ \left[\sum_{k=1}^{\alpha} \mu_y(y_k)\right] \tag{3.14}$$

If α is large then data storage may be a problem for the computation of (3.14). This equation can, however, be evaluated using parallel processing, in this case data storage will not be problem. Currently, however, most researchers still depend on software for simulations and cannot make use of parallel processing. We can, however, use a recursive method to vastly reduce the memory required for storing the data that are needed to compute the defuzzification output. From (3.14), we can calculate

$$A(i) = A(i-1) + y_i \mu_y(y_i) A(0) = 0 \tag{3.15}$$

and

$$B(i) = B(i-1) + y_i \mu_y(y_i) B(0) = 0 \tag{3.16}$$

for $i = 1, \ldots, \alpha$. With these formulas we just need to store A and B during each iteration.

From our previous discussions about the five elements that comprise the Fig. 3.9 type-2 FLS, we see that there are many possibilities to choose from, even more than for a type-1 FLS. To begin, we must decide on the kind of defuzzification (singleton or non-singleton). We must also choose a FOU for each type-2 membership function, decide on the functional forms for both the primary and secondary membership functions, and choose the parameters of the membership functions (fixed a-priori or tuned during a training procedure). Then we need to choose the composition (max-min, max-product), implication (minimum, product), type-reduction method (centroid, center-of-sums, height, modified height, center-of-sets), and defuzzifier. Clearly, there is an even greater richness among type-2 FLSs than there is among type-1 FLSs. In other words, there are more design degrees of freedom associated with a type-2 FLS than with a type-1 FLS; hence, a type-2 FLS has the potential to outperform a type-1 FLS because of the extra degrees of freedom.

3.3.2 Non-Singleton Fuzzy Logic Systems

A non-singleton FLS is one whose inputs are modeled as fuzzy numbers. A type-2 FLS whose inputs are modeled as type-1 fuzzy numbers is referred to as "type-1 non-singleton type-2 FLS". This kind of a fuzzy system not only accounts for uncertainties about either the antecedents or consequents in rules, but also accounts for input measurement uncertainties.

A type-1 non-singleton type-2 FLS is described by the same diagram as in singleton type-2 FLS, see Fig. 3.9. The rules of a type-1 non-singleton type-2 FLS are the same as those for the singleton type-2 FLS. What are different is the fuzzifier, which treats the inputs as type-1 fuzzy sets, and the effect of this on the inference block. The output of the inference block will again be a type-2 fuzzy set; so, the type-reducers and defuzzifier that we described for a singleton type-2 FLS apply as well to a type-1 non-singleton type-2 FLS.

We can also have a situation in which the input are modeled as type-2 fuzzy numbers. This situation can occur, e.g., in time series forecasting when the additive measurement noise is non-stationary. A type-2 FLS whose inputs are modeled as type-2 fuzzy numbers is referred to as "type-2 non-singleton type-2 FLS".

A type-2 non-singleton type-2 FLS is described by the same diagram as in singleton type-2 FLS, see Fig. 3.9. The rules of a type-2 non-singleton type-2 FLS are the same as those for a type-1 non-singleton type-2 FLS, which are the same as those for a singleton type-2 FLS. What is different is the fuzzifier, which treats the inputs as type-2 fuzzy sets, and the effect of this on the inference block. The output of the inference block will again be a type-2 fuzzy set; so, the type-reducers and defuzzifier that we described for a type-1 non-singleton type-2 FLS apply as well to a type-2 non-singleton type-2 FLS.

3.3.3 Sugeno Type-2 Fuzzy Systems

All of our previous FLSs were of the Mamdani type, even though we did not refer to them as such. In this section, we will need to distinguish between the two kinds of FLSs, we refer to our previous FLSs as "Mamdani" FLSs. Both kinds of FLS are characterized by if-then rules and have the same antecedent structures. They differ in the structures of their consequents. The consequent of a Mamdani rule is a fuzzy set, while the consequent of a Sugeno rule is a function.

A type-1 Sugeno FLS was proposed by Takagi & Sugeno (1985), and Sugeno & Kang (1988), in an effort to develop a systematic approach to generating fuzzy rules from a given input-output data set. We will consider in this section the extension of first-order type-1 Sugeno FLS to its type-2 counterpart, with emphasis on interval sets.

Consider a type-2 Sugeno FLS having r inputs $x_1 \in X_1, \ldots, x_r \in X_r$ and one output $y \in Y$. A type-2 Sugeno FLS is also described by fuzzy if-then rules that represent input-output relations of a system. In a general first-order type-2 Sugeno model with a rule base of M rules, each having r antecedents, the ith rule can be expressed as

$$R^l : \text{ IF } x_1 \text{ is } \tilde{A}_1^l \text{ and } \ldots \ x_p \text{ is } \tilde{A}_p^l, \text{ THEN } Y^i = C_0^i + C_1^i x_1 + \cdots + C_r^i x_r$$

$$(3.17)$$

where $i = 1, \ldots, M; C_j^i (j = 1, \ldots, r)$ are consequent type-1 fuzzy sets; Y^i, the output of the ith rule, is also a type-1 fuzzy set (because it is a linear combination of type-1 fuzzy sets); and $\tilde{A}_k^i$ ($k = 1, \ldots, r$) are type-2 antecedent fuzzy sets. These rules let us simultaneously account for uncertainty about antecedent membership functions and consequent parameter values. For a type-2 Sugeno FLS there is no need for type-reduction, just as there is no need for defuzzification in a type-1 Sugeno FLS.

3.4 Introduction to Intuitionistic Fuzzy Logic

The intuitionistic fuzzy sets where defined as an extension of the ordinary fuzzy sets (Atanassov, 1999). As opposed to a fuzzy set in X (Zadeh, 1971), given by

$$B = \{(x, \mu_B(x)) \mid x \in X\} \tag{3.18}$$

where $\mu_B : X \to [0, 1]$ is the membership function of the fuzzy set B, an intuitionistic fuzzy set A is given by

$$A = \{(x, \mu_A(x), \nu_A(x)) \mid x \in X\} \tag{3.19}$$

where $\mu_A : X \to [0, 1]$ and $\nu_A : X \to [0, 1]$ are such that

$$0 \leq \mu_A + \nu_A \leq 1 \tag{3.20}$$

and $\mu_A(x); \nu_A(x) \in [0, 1]$ denote a degree of membership and a degree of non-membership of $x \in A$, respectively.

For each intuitionistic fuzzy set in X, we will call

$$\pi_A(x) = 1 - \mu_A(x) - \nu_A(x) \tag{3.21}$$

a "hesitation margin" (or an "intuitionistic fuzzy index") of $x \in A$ and, it expresses a hesitation degree of whether x belongs to A or not. It is obvious that $0 \leq \pi_A(x) \leq 1$, for each $x \in X$.

On the other hand, for each fuzzy set B in X, we evidently have that

$$\pi_B(x) = 1 - \mu_B(x) - [1 - \mu_B(x)] = 0 \text{ for each } x \in X . \tag{3.22}$$

Therefore, if we want to fully describe an intuitionistic fuzzy set, we must use any two functions from the triplet (Szmidt & Kacprzyk, 2002):

- Membership function,
- Non-membership function,
- Hesitation margin.

In other words, the application of intuitionistic fuzzy sets instead of fuzzy sets means the introduction of another degree of freedom into a set description (i.e. in addition to μ_A we also have ν_A or π_A).

Since the intuitionistic fuzzy sets being a generalization of fuzzy sets give us an additional possibility to represent imperfect knowledge, they can make it possible to describe many real problems in a more adequate way.

Basically, intuitionistic fuzzy sets based models maybe adequate in situations when we face human testimonies, opinions, etc. involving two (or more) answers of the type (Szmidt & Kacprzyk, 2002):

- Yes,
- No,
- I do not know, I am not sure, etc.

Voting can be a good example of such a situation, as human voters may be divided into three groups of those who:

- Vote for,
- Vote against,
- Abstain or give invalid votes.

This third group is of great interest from the point of view of, say, customer behavior analysis, voter behavior analysis, etc., because people from this third undecided group after proper enhancement (eg., different marketing activities) can finally become sure, i.e. become persons voting for (or against), customers wishing to buy products advertised, etc.

3.5 Intuitionistic Fuzzy Inference Systems

Fuzzy inference in intuitionistic systems has to consider the fact that we have the membership μ functions as well as the non-membership ν functions. In this case, we propose that the conclusion of an intuitionistic fuzzy system can be a linear combination of the results of two classical fuzzy systems, one for the membership functions and the other for the non-membership functions.

Assume that IFS is the output of an intuitionistic fuzzy system, then with the following equation we can calculate the total output as a linear combination:

$$\text{IFS} = (1 - \pi)\text{FS}_\mu + \pi\text{FS}_\nu \qquad (3.23)$$

where FS_μ is the traditional output of a fuzzy system using the membership function μ, and FS_ν is the output of a fuzzy system using the non-membership function ν. Of course (3.23) for $\pi = 0$ will reduce to the output of a traditional fuzzy system, but for other values of π the result of IFS will be different as we are now taking into account the hesitation margin π.

The advantage of this method for computing the output IFS of an intuitionistic fuzzy system is that we can use our previous machinery of traditional fuzzy systems for computing FS_μ and FS_ν. Then, we only perform a weighted average of both results to obtain the final output IFS of the intuitionistic fuzzy inference system. We consider below a simple example to illustrate these ideas.

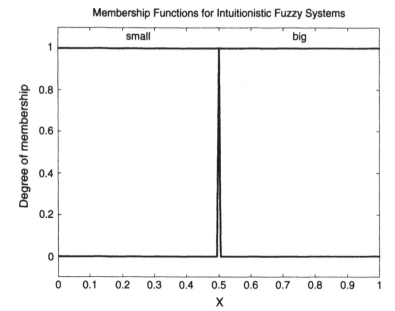

Fig. 3.11. Membership functions for the "small" and "big" linguistic values of the input variable

Example 3.6. : Let us assume that we have a simple intuitionistic fuzzy system of only two rules:

$$\text{IF } x \text{ is small} \quad \text{THEN } y \text{ is big}$$
$$\text{IF } x \text{ is big} \quad \text{THEN } y \text{ is small}$$

We will consider for simplicity uniform rectangular membership functions for both linguistic variables. We show in Fig. 3.11 the membership functions for the linguistic values "small" and "big" of the input linguistic variable. We also show in Fig. 3.12 the non-membership functions for the linguistic values of the output variable. It is clear from Fig. 3.12 that in this case the membership and non-membership functions are not complementary, which is due to the fact that we have an intuitionistic fuzzy system.

From Fig. 3.12 we can clearly see that the hesitation margin π is 0.05 for both cases. As a consequence (3.22) can be written for our example as follows:

$$\text{IFS} = 0.95\text{FS}_\mu + 0.05\text{FS}_\nu \tag{3.24}$$

Now, let us assume that we want to calculate the output of the fuzzy system for a given input value of $x = 0.45$. In this case, we have that x is small with $\mu = 1$ and x is not small with $\nu = 0$, and x is big with $\mu = 0$ and x is not big with $\nu = 0.5$. As a consequence of these facts we have that,

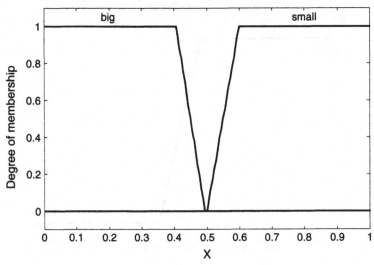

Fig. 3.12. Non-membership functions for the "small" and "big" linguistic values of the output variable

$$\text{IFS} = \text{IFS}_{\text{small}} + \text{IFS}_{\text{big}}$$
$$\text{IFS} = 0.95\text{FS}_{\mu\,\text{small}} + 0.05\text{FS}_{\nu\,\text{small}} + 0.95\text{FS}_{\mu\,\text{big}} + 0.05\text{FS}_{\nu\,\text{big}}$$
$$\text{IFS} = 0.95\text{FS}_{\mu\,\text{small}} + 0.05\text{FS}_{\nu\,\text{big}}$$
$$\text{IFS} = 0.95(0.75) + 0.05(0.765)$$
$$\text{IFS} = 0.74075$$

Of course, we can compare this intuitionistic fuzzy output with the traditional one (of 0.75), the difference between these two output values is due to the hesitation margin. We have to mention that in this example the difference is small because the hesitation margin is also small. We show in Table 3.1 the results of the intuitionistic fuzzy system for several input values.

We can appreciate from Table 3.1 the difference between the outputs of the intuitionistic fuzzy system and the output of the classical one.

Table 3.1. Sample results of the intuitionistic fuzzy system for several input values

Input Values, x	Membership Result	Non-Membership Result	Intuitionistic Result
0.2500	0.7500	0.7766	0.741330
0.3500	0.7500	0.7766	0.741330
0.4500	0.7500	0.7650	0.740750
0.5500	0.2500	0.2359	0.249295
0.6500	0.2500	0.2250	0.248750
0.7500	0.2500	0.2250	0.248750

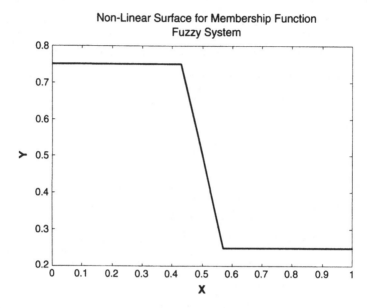

Fig. 3.13. Non-linear surface of the membership function fuzzy system

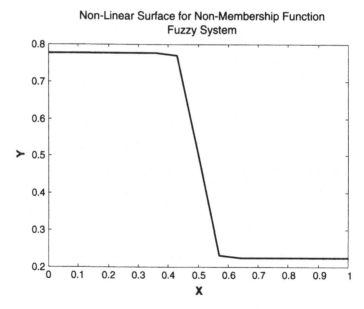

Fig. 3.14. Non-linear surface of non-membership function fuzzy system

Finally, we show in Figs. 3.13 and 3.14 the non-linear surfaces for the fuzzy systems of the membership and non-membership functions, respectively. We can appreciate from these figures that the surfaces are similar, but they differ slightly because of the hesitation margin.

3.6 Summary

In this chapter, we have presented the main ideas underlying intuitionistic and type-2 fuzzy logic and we have only started to point out the many possible applications of these powerful computational theories. We have discussed in some detail type-2 fuzzy set theory, fuzzy reasoning and fuzzy inference systems. At the end, we also gave some remarks about type-2 fuzzy modeling with the Mamdani and Sugeno approaches. We have also discussed in some detail intuitionistic fuzzy sets and intuitionistic fuzzy inference systems. In the following chapters, we will show how intuitionistic and type-2 fuzzy logic (in some cases, in conjunction with other methodologies) can be applied to solve real world complex problems. This chapter will serve as a basis for the new hybrid intelligent methods, for modeling, simulation, and pattern recognition that will be described later this book.

4

Supervised Learning Neural Networks

In this chapter, we describe the basic concepts, notation, and basic learning algorithms for supervised neural networks that will be of great use for solving pattern recognition problems in the following chapters of this book. The chapter is organized as follows: backpropagation for feedforward networks, radial basis networks, adaptive neuro-fuzzy inference systems (ANFIS) and applications. First, we give a brief review of the basic concepts of neural networks and the basic backpropagation learning algorithm. Second, we give a brief description of the momentum and adaptive momentum learning algorithms. Third, we give a brief review of the radial basis neural networks. Finally, we end the chapter with a description of the adaptive neuro-fuzzy inference system (ANFIS) methodology. We consider this material necessary to understand the new methods for pattern recognition that will be presented in the final chapters of this book.

4.1 Backpropagation for Feedforward Networks

This section describes the architectures and learning algorithms for adaptive networks, a unifying framework that subsumes almost all kinds of neural network paradigms with supervised learning capabilities. An adaptive network, as the name indicates, is a network structure consisting of a number of nodes connected through directional links. Each node represents a process unit, and the links between nodes specify the causal relationship between the connected nodes. The learning rule specifies how the parameters (of the nodes) should be updated to minimize a prescribed error measure.

The basic learning rule of the adaptive network is the well-known steepest descent method, in which the gradient vector is derived by successive invocations of the chain rule. This method for systematic calculation of the gradient vector was proposed independently several times, by Bryson and Ho (1969), Werbos (1974), and Parker (1982). However, because research on artificial neural networks was still in its infancy at those times, these researchers' early

Patricia Melin and Oscar Castillo: *Hybrid Intelligent Systems for Pattern Recognition Using Soft Computing*, StudFuzz **172**, 55–83 (2005)
www.springerlink.com

work failed to receive the attention it deserved. In 1986, Rumelhart et al.
used the same procedure to find the gradient in a multilayer neural network.
Their procedure was called "backpropagation learning rule", a name which
is now widely known because the work of Rumelhart inspired enormous interest in research on neural networks. In this section, we introduce Werbos's
original backpropagation method for finding gradient vectors and also present
improved versions of this method.

4.1.1 The Backpropagation Learning Algorithm

The procedure for finding a gradient vector in a network structure is generally
referred to as "backpropagation" because the gradient vector is calculated
in the direction opposite to the flow of the output of each node. Once the
gradient is obtained, a number of derivative-based optimization and regression
techniques are available for updating the parameters. In particular, if we use
the gradient vector in a simple steepest descent method, the resulting learning
paradigm is referred to as "backpropagation learning rule". Suppose that a
given feedforward adaptive network has L layers and layer l $(l = 0, 1, \ldots, L)$
has $N(l)$ nodes. Then the output and function of node i $[i = 1, \ldots, N(l)]$ in
layer l can be represented as $x_{l,i}$ and $f_{l,i}$, respectively, as shown in Fig. 4.1.
Since the output of a node depends on the incoming signals and the parameter
set of the node, we have the following general expression for the node function
$f_{l,i}$:

$$X_{l,i} = f_{l,i}(x_{i-1,1}, \ldots, x_{l-1,N(l-1)}, \alpha, \beta, \gamma, \ldots) \tag{4.1}$$

where α, β, γ, etc. are the parameters of this node.

Assuming that the given training data set has P entries, we can define an
error measure for the pth $(1 \le p \le P)$ entry of the training data as the sum
of the squared errors:

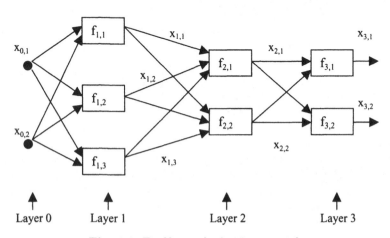

Fig. 4.1. Feedforward adaptive network

$$E_p = \sum_{k=1}^{N(L)} (d_k - x_{L,k})^2 \qquad (4.2)$$

where d_k is the kth component of the pth desired output vector and $x_{L,k}$ is the kth component of the actual output vector produced by presenting the pth input vector to the network. Obviously, when E_p is equal to zero, the network is able to reproduce exactly the desired output vector in the pth training data pair. Thus our task here is to minimize an overall error measure, which is defined as $E = \sum E_p$.

We can also define the "error signal" $\varepsilon_{l,i}$ as the derivative of the error measure E_p with respect to the output of the node i in layer l, taking both direct and indirect paths into consideration. Mathematically,

$$\varepsilon_{l,i} = \frac{\partial^+ E_p}{\partial x_{l,i}} \qquad (4.3)$$

this expression was called the "ordered derivative" by Werbos (1974). The difference between the ordered derivative and the ordinary partial derivative lies in the way we view the function to be differentiated. For an internal node output $x_{l,i}$, the partial derivative $\partial^+ E_p / \partial x_{l,i}$ is equal to zero, since E_p does not depend on $x_{l,i}$ directly. However, it is obvious that E_p does depend on $x_{l,i}$ indirectly, since a change in $x_{l,i}$ will propagate through indirect paths to the output layer and thus produce a corresponding change in the value of E_p.

The error signal for the ith output node (at layer L) can be calculated directly:

$$\varepsilon_{L,i} = \frac{\partial^+ E_p}{\partial x_{L,i}} = \frac{\partial E_p}{\partial x_{L,i}} \qquad (4.4)$$

This is equal to $\varepsilon_{L,i} = -2(d_i - x_{L,i})$ if E_p is defined as in (4.2). For the internal node at the ith position of layer l, the error signal can be derived by the chain rule of differential calculus:

$$\varepsilon_{l,i} = \underbrace{\frac{\partial^+ E_p}{\partial x_{l,i}}}_{\substack{\text{error signal} \\ \text{at layer } l}} = \sum_{m=1}^{N(l+1)} \underbrace{\frac{\partial^+ E_p}{\partial x_{l+1,m}}}_{\substack{\text{error signal} \\ \text{at layer } l+1}} \frac{\partial f_{l+1,m}}{\partial x_{l,i}} = \sum_{m=1}^{M(l+1)} \epsilon_{l+1,m} \frac{\partial f_{l+1,m}}{\partial x_{l,i}} \qquad (4.5)$$

where $0 \le l \le L - 1$. That is, the error signal of an internal node at layer l can be expressed as a linear combination of the error signal of the nodes at layer $l + 1$. Therefore, for any l and i, we can find $\varepsilon_{l,i}$ by first applying (4.4) once to get error signals at the output layer, and then applying (4.5) iteratively until we reach the desired layer l. The underlying procedure is called backpropagation since the error signals are obtained sequentially from the output layer back to the input layer.

The gradient vector is defined as the derivative of the error measure with respect to each parameter, so we have to apply the chain rule again to find the gradient vector. If α is a parameter of the ith node at layer l, we have

$$\frac{\partial^+ E_p}{\partial \alpha} = \frac{\partial^+ E_p}{\partial x_{l,i}} \frac{\partial f_{l,i}}{\partial \alpha} = \varepsilon_{l,i} \frac{\partial f_{l,i}}{\partial \alpha} \tag{4.6}$$

The derivative of the overall error measure E with respect to α is

$$\frac{\partial^+ E}{\partial \alpha} = \sum_{p=1}^{p} \frac{\partial^+ E_p}{\partial \alpha} \tag{4.7}$$

Accordingly, for simple steepest descent (for minimization), the update formula for the generic parameter α is

$$\Delta \alpha = -\eta \frac{\partial^+ E}{\partial \alpha} \tag{4.8}$$

in which η is the "learning rate", which can be further expressed as

$$\eta = \frac{k}{\sqrt{\sum_{\alpha} (\partial E / \partial \alpha)^2}} \tag{4.9}$$

where k is the "step size", the length of each transition along the gradient direction in the parameter space.

There are two types of learning paradigms that are available to suit the needs for various applications. In "off-line learning" (or "batch learning"), the update formula for parameter α is based on (4.7) and the update action takes place only after the whole training data set has been presented-that is, only after each "epoch" or "sweep". On the other hand, in "on-line learning" (or "pattern-by-pattern learning"), the parameters are updated immediately after each input-output pair has been presented, and the update formula is based on (4.6). In practice, it is possible to combine these two learning modes and update the parameter after k training data entries have been presented, where k is between 1 and P and it is sometimes referred to as the "epoch size".

4.1.2 Backpropagation Multilayer Perceptrons

Artificial neural networks, or simply "neural networks" (NNs), have been studied for more than three decades since Rosenblatt first applied single-layer "perceptrons" to pattern classification learning (Rosenblatt, 1962). However, because Minsky and Papert pointed out that single-layer systems were limited and expressed pessimism over multilayer systems, interest in NNs dwindled in the 1970s (Minsky & Papert, 1969). The recent resurgence of interest in the field of NNs has been inspired by new developments in NN learning algorithms (Fahlman & Lebiere, 1990), analog VLSI circuits, and parallel processing techniques (Lippmann, 1987).

Quite a few NN models have been proposed and investigated in recent years. These NN models can be classified according to various criteria, such as

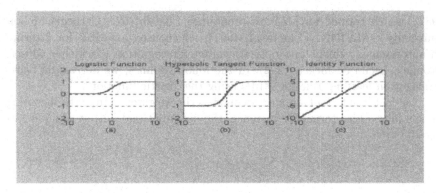

Fig. 4.2. Activation functions for backpropagation MLPs: (**a**) logistic function; (**b**) hyperbolic function; (**c**) identity function

their learning methods (supervised versus unsupervised), architectures (feed-forward versus recurrent), output types (binary versus continuous), and so on. In this section, we confine our scope to modeling problems with desired input-output data sets, so the resulting networks must have adjustable parameters that are updated by a supervised learning rule. Such networks are often referred to as "supervised learning" or "mapping networks", since we are interested in shaping the input-output mappings of the network according to a given training data set.

A backpropagation "multilayer perceptron" (MLP) is an adaptive network whose nodes (or neurons) perform the same function on incoming signals; this node function is usually a composite of the weighted sum and a differentiable non-linear activation function, also known as the "transfer function". Figure 4.2 depicts three of the most commonly used activation functions in backpropagation MLPs:

Logistic function: $\qquad\qquad\qquad\qquad f(x) = \dfrac{1}{1 + e^{-x}}$

Hyperbolic tangent function: $\quad f(x) = \tanh(x/2) = \dfrac{1 - e^{-x}}{1 + e^{-x}}$

Identity function: $\qquad\qquad\qquad\qquad f(x) = x$

Both the hyperbolic tangent and logistic functions approximate the signum and step function, respectively, and provide smooth, nonzero derivatives with respect to input signals. Sometimes these two activation functions are referred to as "squashing functions" since the inputs to these functions are squashed to the range $[0, 1]$ or $[-1, 1]$. They are also called "sigmoidal functions" because their s-shaped curves exhibit smoothness and asymptotic properties.

Backpropagation MLPs are by far the most commonly used NN structures for applications in a wide range of areas, such as pattern recognition, signal processing, data compression and automatic control. Some of the well-known instances of applications include NETtalk (Sejnowski & Rosenberg,

1987), which trained an MLP to pronounce English text, Carnegie Mellon University's ALVINN (Pomerleau, 1991), which used an MLP for steering an autonomous vehicle; and optical character recognition (Sackinger, Boser, Bromley, Lecun & Jackel, 1992). In the following lines, we derive the back-propagation learning rule for MLPs using the logistic function.

The "net input" x of a node is defined as the weighted-sum of the incoming signals plus a bias term. For instance, the net input and output of node j in Fig. 4.3 are

$$x_j = \sum_i w_{ij} x_i + w_j \ ,$$

$$x_j = f(x_j) = \frac{1}{1 + \exp(-x_j)} \ , \qquad (4.10)$$

where x_i is the output of node i located in any one of the previous layers, w_{ij} is the weight associated with the link connecting nodes i and j, and w_j is the bias of node j. Since the weights w_{ij} are actually internal parameters associated with each node j, changing the weights of a node will alter the behavior of the node and in turn alter the behavior of the whole backpropagation MLP.

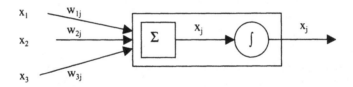

Fig. 4.3. Node j of a backpropagation MLP

Figure 4.4 shows a three-layer backpropagation MLP with three inputs to the input layer, three neurons in the hidden layer, and two output neurons in the output layer. For simplicity, this MLP will be referred to as a 3-3-2 network, corresponding to the number of nodes in each layer.

The "backward error propagation", also known as the "backpropagation" (BP) or the "generalized data rule" (GDR), is explained next. First, a squared error measure for the pth input-output pair is defined as

$$E_p = \sum_k (d_k - x_k)^2 \qquad (4.11)$$

where d_k is the desired output for node k, and x_k is the actual output for node k when the input part of the pth data pair presented. To find the gradient vector, an error term ε_i is defined as

$$\varepsilon_i = \frac{\partial^+ E_p}{\partial x_i} \qquad (4.12)$$

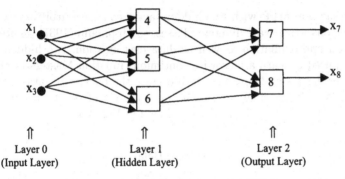

Fig. 4.4. A 3-3-2 backpropagation MLP

By the chain rule, the recursive formula for ε_i can be written as

$$\varepsilon_i = \begin{cases} -2(d_i - x_i)\frac{\partial x_i}{\partial x_i} = -2(d_i - x_i)x_i(1 - x_i) & \text{if node } i \text{ is a} \\ & \text{output node} \\ \frac{\partial x_i}{\partial x_i}\sum_{j,i<j}\frac{\partial^+ E_p}{\partial x_j}\frac{\partial x_j}{\partial x_i} = x_i(1 - x_i)\sum_{j,i<j}\varepsilon_j w_{ij} & \text{otherwise} \end{cases} \quad (4.13)$$

where w_{ij} is the connection weight from node i to j; and w_{ij} is zero if there is no direct connection. Then the weight update w_{ki} for on-line (pattern-by-pattern) learning is

$$\Delta w_{ki} = -\eta\frac{\partial^+ E_p}{\partial w_{ki}} = -\eta\frac{\partial^+ E_p}{\partial x_i}\frac{\partial x_i}{\partial w_{ki}} = -\eta\varepsilon_i x_k \quad (4.14)$$

where η is a learning rate that affects the convergence speed and stability of the weights during learning.

For off-line (batch) learning, the connection weight w_{ki} is updated only after presentation of the entire data set, or only after an "epoch":

$$\Delta w_{ki} = -\eta\frac{\partial^+ E}{\partial w_{ki}} = -\eta\sum_p\frac{\partial^+ E_p}{\partial w_{ki}} \quad (4.15)$$

or, in vector form,

$$\Delta\mathbf{w} = -\eta\frac{\partial^+ E}{\partial\mathbf{w}} = -\eta\nabla_{\mathbf{w}}E \quad (4.16)$$

where $E = \sum_p E_p$. This corresponds to a way of using the true gradient direction based on the entire data set.

The approximation power of backpropagation MLPs has been explored by some researchers. Yet there is very little theoretical guidance for determining network size in terms of say, the number of hidden nodes and hidden layers it should contain. Cybenko (1989) showed that a backpropagation MLP, with one hidden layer and any fixed continuous sigmoidal non-linear function, can approximate any continuous function arbitrarily well on a compact set. When used as a binary-valued neural network with the step activation function, a backpropagation MLP with two hidden layers can form arbitrary complex decision regions to separate different classes, as Lippmann (1987) pointed out. For function approximation as well as data classification, two hidden layers may be required to learn a piecewise-continuous function Masters (1993).

Lets consider a simple example to illustrate the backpropagation learning algorithm. We will consider as training data the set of points distributed in the square $[-1, 1] \times [-1, 1]$, which are shown in Fig. 4.5.

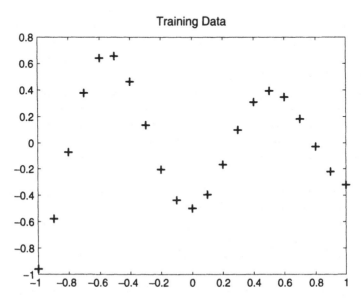

Fig. 4.5. Training data for the backpropagation learning algorithm

We show in Fig. 4.6 the initial approximation of a three layer neural network with 10 neurons in the hidden layer and hyperbolic tangent activation functions. The initial approximation is quite bad because the initial weights of the network are generated randomly. After training with the backpropagation algorithm for 1000 epochs with arrive to the final approximation shown in Fig. 4.7, which has a final sum of squares errors SSE = 0.0264283.

4.1.3 Methods for Speeding Up Backpropagation

One way for to speed up backpropagation is to use the so-called "momentum term" (Rumelhart, Hinton & Williams, 1986):

$$\Delta \mathbf{w} = \eta \nabla_{\mathbf{w}} E + \alpha \Delta \mathbf{w}_{\text{prev}} \tag{4.17}$$

where $\mathbf{w}_{\text{prev}}$ is the previous value of the weights, and the "momentum constant" α, in practice, is usually set to a value between 0.1 and 1. The addition

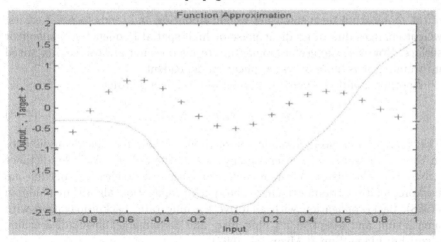

Fig. 4.6. Initial approximation of the backpropagation learning algorithm

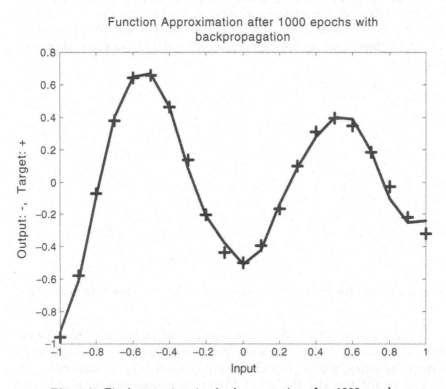

Fig. 4.7. Final approximation backpropagation after 1000 epochs

of the momentum term smoothes weight updating and tends to resist erratic weight changes due to gradient noise or high spatial frequencies in the error surface. However, the use of momentum terms does not always seem to speed up training; it is more or less application dependent.

Another useful technique is normalized weight updating:

$$\Delta \mathbf{w} = -\kappa(\nabla_{\mathbf{w}} E)/(\|\nabla_{\mathbf{w}} E\|) \tag{4.18}$$

This causes the network's weight vector to move the same Euclidean distance in the weight space with each update, which allows control of the distance κ based on the history of error measures. Other strategies for speeding up backpropagation training include the quick-propagation algorithm Fahlman (1988), backpropagation with adaptive learning rate, backpropagation with momentum and adaptive learning rate, and Levenberg-Marquardt learning algorithm (Jang, Sun & Mizutani, 1997).

Lets consider a simple example to illustrate these methods. We will again use as training data the set of points shown in Fig. 4.5. We will first apply the backpropagation with momentum learning algorithm with the same parameters as before. We show in Fig. 4.8 the initial function approximation and in Fig. 4.9 the final function approximation achieved with backprogration with momentum.

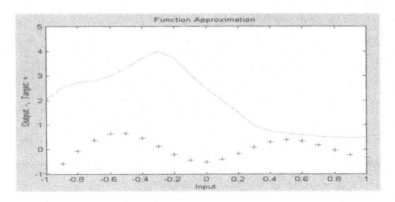

Fig. 4.8. Initial function approximation with backpropagation with momentum

In this case, the final SSE is of 0.0082689, which is lower than the one obtained by simple backpropagation. As a consequence we are achieving a better final approximation with the backpropagation with momentum.

Now we will consider the use of "backpropagation with momentum and adaptive learning rate". In this case, the learning rate is not fixed as in the previous methods, instead it is changed according to the error surface. We will again consider the training data of Fig. 4.5. The initial function approximation is shown in Fig. 4.10. The final function approximation is shown in Fig. 4.11,

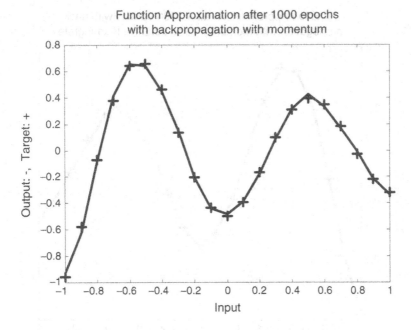

Fig. 4.9. Final function approximation with backpropagation with momentum

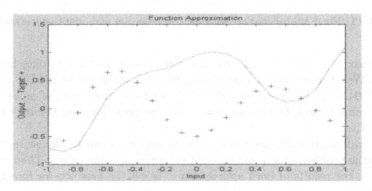

Fig. 4.10. Initial approximation with backpropagation and adaptive learning rate

which is achieved after 1000 epochs with the same network architecture as before.

In this case, the final approximation achieved with the "backpropagation method with adaptive learning rate" is even better because the SSE is of only 0.0045014. This SSE is lower than the ones obtained previously with the other methods.

We will now consider the more complicated problem of forecasting the prices of onion and tomato in the U.S. market. The time series for the prices of these consumer goods show very complicated dynamic behavior, and for

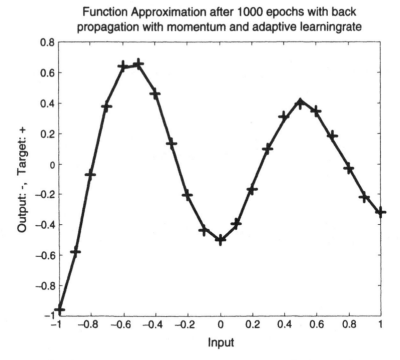

Fig. 4.11. Final approximation with backpropagation and adaptive learning rate

this reason it is interesting to analyze and predict the future prices for these goods. We show in Table 4.1 the time series of onion prices in the U.S. market from 1994 to 2000. We also show in Fig. 4.12 the time series of onion prices in the same period, to give an idea of the complex dynamic behavior of this time series.

We will apply both the neural network approach and the linear regression one to the problem of forecasting the time series of onion prices. Then, we will compare the results of both approaches to select the best one for forecasting.

Table 4.1. Time series of onion prices in the U.S. Market for 1994–2000 period

Month	1994–1995	1995–1996	1996–1997	1997–1998	1998–1999	1999–2000
October	4.13	7.30	6.60	5.63	8.74	6.89
November	4.20	7.50	6.69	5.72	4.66	6.93
December	4.38	7.60	6.30	4.80	8.11	5.63
January	5.16	8.21	6.10	4.75	6.19	3.92
February	5.14	6.00	5.84	4.83	5.97	4.59
March	4.65	7.86	4.73	4.75	4.86	3.76
April	4.89	5.58	5.38	4.75	5.62	3.19

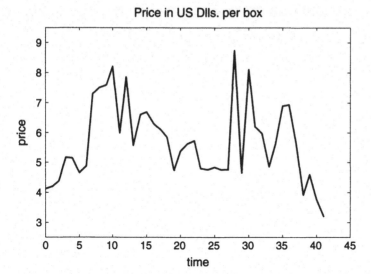

Fig. 4.12. Price in US Dollars per box of onion from October 1994 to April 2000

We also show in Table 4.2 the time series of tomato prices in the U.S. Market from 1994 to 1999.

First, we will describe the results of applying neural networks to the time series of onion prices (Table 4.1). The data given in Table 4.1 was used as training data for feedforward neural networks with two different learning algorithms. The adaptive learning backpropagation with momentum, and the Levenberg-Marquardt training algorithms were used for the neural networks.

We show in Fig. 4.13 the result of training a three layer (85 nodes in the hidden layer) feedforward neural network with the Levenberg-Marquardt learning algorithm. In Fig. 4.13, we can see how the neural network approximates very well the real time series of onion prices over the relevant period of time. Actually, we have seen that the approximating power for the Levenberg-Marquardt learning algorithm is better.

Table 4.2. Time series of tomato prices in the U.S. Market for the 1994–1999 period

Month	1994	1995	1996	1997	1998	1999
June	28.40	32.80	26.80	27.30	25.70	27.80
July	23.30	17.10	23.50	25.40	43.10	20.30
August	27.40	12.70	20.60	25.40	20.40	22.50
September	19.10	17.20	22.40	23.20	26.60	25.30
October	25.70	20.20	27.60	23.30	43.10	18.90
November	28.80	22.70	31.60	41.10	35.80	20.30

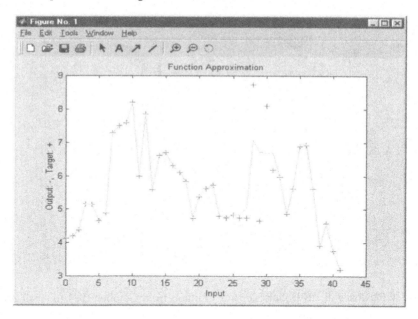

Fig. 4.13. Neural network for onion prices with the Levenberg-Marquardt algorithm

Now we describe the results of applying neural networks to the time series of tomato prices (Table 4.2). We show in Fig. 4.14 the result of training a three layer (85 nodes in hidden layer) neural network with the Levenberg-Marquardt learning algorithm. In Fig. 4.14, we can see how the neural network approximates very well the time series of tomato prices.

We also show in Fig. 4.15 the results of forecasting tomato prices from 2000 to 2010 using the neural network with the Levenberg-Marquardt learning algorithm. Predictions are considered very good by experts at the beginning (first three years or so) but after that they may be not so good. In any case, the results are better than the ones obtained with classical regression methods.

We summarize the above results using neural networks, and the results of using linear regression models in Table 4.3. From Table 4.3 we can see very clearly the advantage of using neural networks for simulating and forecasting the time series of prices. Also, we can conclude from this table that the Levenberg-Marquardt training algorithm is better than backpropagation with momentum. Of course, the reason for this may be that the Levenberg-Marquardt training algorithm, having a variable learning rate is able to adjust to the complicated behavior of the time series. Finally, we have to mention that the problem of time series analysis can be considered as one of pattern recognition, because we are basically finding or learning the patterns in data. For this reason, it is very important to be able to use neural networks in this type of problems. We will concentrate in this book more on pattern

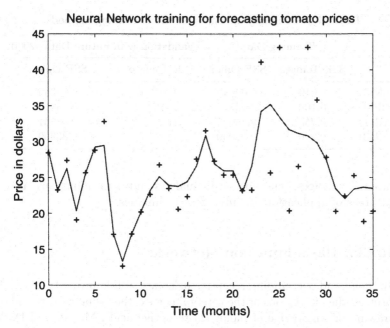

Fig. 4.14. Neural network for tomato prices with the Levenberg-Marquardt algorithm

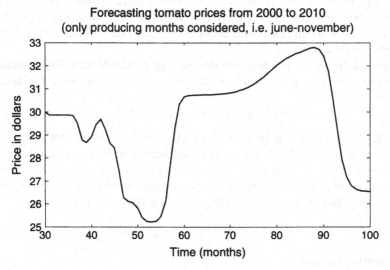

Fig. 4.15. Forecasting tomato prices from 2000 to 2010 with the neural network

Table 4.3. Summary of results for the forecasting methods

	Training Data		Validation with Future Data (2000)	
	SSE Tomato	SSE Onion	SSE Tomato	SSE Onion
NN LM	0.019	0.00039	0.025	0.00055
NN BPM	0.100	0.00219	0.150	0.00190
LR AR(1)	0.250	0.05000	0.551	0.08501
LR AR(2)	0.200	0.01000	0.511	0.03300

recognition for images, but time series analysis and data mining are also important areas of application for intelligent techniques.

4.2 Radial Basis Function Networks

Locally tuned and overlapping receptive fields are well-known structures that have been studied in regions of the cerebral cortex, the visual cortex, and others. Drawing on knowledge of biological receptive fields, Moody and Darken (1989) proposed a network structure that employs local receptive fields to perform function mappings. Similar schemes have been proposed by Powell (1987) and many others in the areas of "interpolation" and "approximation theory"; these schemes are collectively call radial basis function approximations. Here we will call the neural network structure the "radial basis function network" or RBFN.

Figure 4.16 shows a schematic diagram of a RBFN with four receptive field units; the activation level of the ith receptive field unit (or hidden unit) is

$$w_i = R_i(\mathbf{x}) = R_i(||\mathbf{x} - \mathbf{u}_i||/\sigma_i), i = 1, 2, \ldots, H , \qquad (4.19)$$

where $\mathbf{x}$ is a multidimensional input vector, $\mathbf{u}_i$ is a vector with the same dimension as $\mathbf{x}$, H is the number of radial basis functions (or, equivalently, receptive field units), and $R_i(\ )$ is the ith radial basis function with a single maximum at the origin. There are no connection weights between the input layer and the hidden layer. Typically, $R_i(\ )$ is a Gaussian function

$$R_i(\mathbf{x}) = \exp\left[-(||\mathbf{x} - \mathbf{u}_i||^2)/2\sigma_i^2\right] \qquad (4.20)$$

or a logistic function

$$R_i(\mathbf{x}) = 1/[1 + \exp\left[(||\mathbf{x} - \mathbf{u}_i||^2)/\sigma_i^2\right] \qquad (4.21)$$

Thus, the activation level of radial basis function wi computed by the ith hidden unit is maximum when the input vector $\mathbf{x}$ is at the center $\mathbf{u}_i$ of that unit.

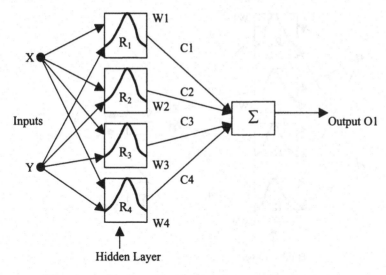

Fig. 4.16. Single-output RBFN that uses weighted sum

The output of an RBFN can be computed in two ways. In the simpler method, as shown in Fig. 4.12, the final output is the weighted sum of the output value associated with each receptive field:

$$d(\mathbf{x}) = \sum_{i=1}^{H} c_i w_i = \sum_{i=1}^{H} c_i R_i(\mathbf{x}) \tag{4.22}$$

where c_i is the output value associated with the ith receptive field. We can also view c_i as the connection weight between the receptive field i and the ouput unit. A more complicated method for calculating the overall output is to take the weighted average of the output associated with each receptive field:

$$d(\mathbf{x}) = \left(\sum_{i=1}^{H} c_i w_i \right) \Big/ \left(\sum_{i=1}^{H} w_i \right) \tag{4.23}$$

Weighted average has a higher degree of computational complexity, but it has the advantage that points in the areas of overlap between two or more receptive fields will have a well-interpolated overall output between the outputs of the overlapping receptive fields.

For representation purposes, if we change the radial basis function $R_i(x)$ in each node of layer 2 in Fig. 4.16 to its normalized counterpart $R_i(x)/\sum_i R_i(x)$, then the overall output is specified by (4.23). A more explicit representation is the shown in Fig. 4.17, where the division of the weighted sum $(\sum_{i=1} c_i\, w_i)$ by the activation total $(\sum_{i=1} w_i)$ is indicated in the division node in the last layer. Of course, similar figures can be drawn for two inputs or more in a RBFN network. We can appreciate from these figures the architecture of this

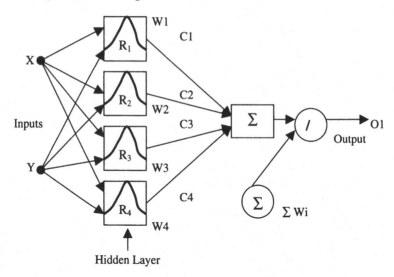

Fig. 4.17. Single-output RBFN that uses weighted average

type of neural networks. As a consequence we can see the difference between RBFN neural networks and MLP networks.

Moody-Darken's RBFN may be extended by assigning a linear function to the output function of each receptive field-that is, making c_i a linear combination of the input variables plus a constant:

$$c_i = \mathbf{a}_i^T \mathbf{x} + b_i \qquad (4.24)$$

where $\mathbf{a}_i$ is a parameter vector and b_i is a scalar parameter. An RBFN's approximation capacity may be further improved with supervised adjustments of the center and shape of the receptive field (or radial basis) functions (Lee & Kil, 1991). Several learning algorithms have been proposed to identify the parameters ($\mathbf{u}_i$, σ_i, and c_i) of an RBFN. Besides using a supervised learning scheme alone to update all modifiable parameters, a variety of sequential training algorithms for RBFNs have been reported. The receptive field functions are first fixed, and then the weights of the output layer are adjusted. Several schemes have been proposed to determine the center positions ($\mathbf{u}_i$) of the receptive field functions. Lowe (1989) proposed a way to determine the centers based on standard deviations of the training data. Moody and Darken (1989) selected the centers $\mathbf{u}_i$ by means of data clustering techniques that assume that similar input vectors produce similar outputs; σ_i's are then obtained heuristically by taking the average distance to the several nearest neighbors of u_i's. Once the non-linear parameters are fixed and the receptive fields are frozen, the linear parameters (i.e., the weights of the output layer) can be updated using either the least squares method or the gradient method.

Using (4.24), extended RBFN response is identical to the response produced by the first-order Sugeno (type-1) fuzzy inference system described in

Chap. 2, provided that the membership functions, the radial basis functions, and certain operators are chosen correctly. While the RBFN consists of radial basis functions, the Sugeno fuzzy system contains a certain number of membership functions. Although the fuzzy system and the RBFN were developed on different bases, they are essentially rooted in the same grounds. Just as the RBFN enjoys quick convergence, the fuzzy system can evolve to recognize some features in a training data set.

Assuming that there is no noise in the training data set, we need to estimate a function $d(.)$ that yields exact desired outputs for all training data. This task is usually called an "interpolation" problem, and the resulting function $d(.)$ should pass through all of the training data points. When we use an RBFN with the same number of basis functions as we have training patterns, we have a so-called "interpolation RBFN", where each neuron in the hidden layer responds to one particular training input pattern.

Lets consider application of the RBFN network to the same example of Fig. 4.5. We will use a two layer RBFN network with 3 neurons in the hidden layer and weighted sum to calculate the output. We show in Fig. 4.18 the Gaussian radial basis function used in the network. Figure 4.19 illustrates the application of weighted sum to achieve the approximation of the training data. Figure 4.20 shows the final approximation achieved with the RBFN network, which is very good. The final SSE is of only 0.002, which is smaller than the one obtained by any of the previous methods. We can conclude that the RBFN network gives the best approximation to the training data of Fig. 4.5.

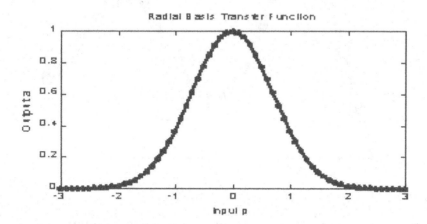

Fig. 4.18. Gaussian radial basis function

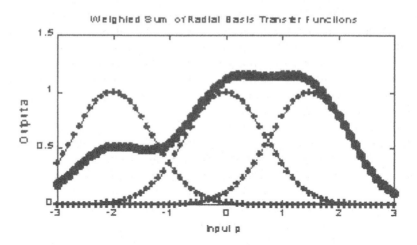

Fig. 4.19. Weighted sum of the three Gaussian functions of the RBFN network

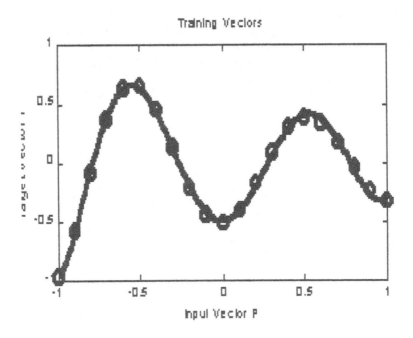

Fig. 4.20. Final function approximation achieved with the RBFN network

4.3 Adaptive Neuro-Fuzzy Inference Systems

In this section, we describe a class of adaptive networks that are functionally equivalent to fuzzy inference systems (Kosko, 1992). The architecture is referred to as ANFIS, which stands for "adaptive network-based fuzzy inference system". We describe how to decompose the parameter set to facilitate the hybrid learning rule for ANFIS architectures representing both the Sugeno and Tsukamoto fuzzy models.

4.3.1 ANFIS Architecture

A fuzzy inference system consists of three conceptual components: a fuzzy rule base, which contains a set of fuzzy if-then rules; a database, which defines the membership functions used in the fuzzy rules; and a reasoning mechanism, which performs the inference procedure upon the rules to derive a reasonable output or conclusion (Kandel, 1992). For simplicity, we assume that the fuzzy inference system under consideration has two inputs x and y and one output z. For a first-order Sugeno fuzzy model (Sugeno & Kang, 1988), a common rule set with two fuzzy if-then rules is the following:

Rule 1: If x is A_1 and y is B_1, then $f_1 = p_1 x + q_1 y + r_1$,

Rule 2: If x is A_2 and y is B_2, then $f_2 = p_2 x + q_2 y + r_2$,

Figure 4.21(a) illustrates the reasoning mechanism for this Sugeno model; the corresponding equivalent ANFIS architecture is as shown in Fig. 4.21(b), where nodes of the same layer have similar functions, as described next. (Here we denote the output of the ith node in layer l as $0_{l,i}$).

Layer 1: Every node i in this layer is an adaptive node with a node function

$$0_{l,i} = \mu_{Ai}(x), \text{ for } i = 1, 2 ,$$

$$0_{l,i} = \mu_{Bi-2}(y), \text{ for } i = 3, 4 , \tag{4.25}$$

where x (or y) is the input to node i and A_i (or B_{i-2}) is a linguistic label (such as "small" or "large") associated with this node. In other words, $0_{l,i}$ is the membership grade of a fuzzy set A and it specifies the degree to which the given input x (or y) satisfies the quantifier A. Here the membership function for A can be any appropriate parameterized membership function, such as the generalized bell function:

$$\mu_A(x) = \frac{1}{1 + |(x - c_i)/a_i|^{2bi}} \tag{4.26}$$

where $\{a_i, b_i, c_i\}$ is the parameter set. As the values of these parameters change, the bell-shaped function varies accordingly, thus exhibiting various forms of membership functions for a fuzzy set A. Parameters in this layer are referred to as "premise parameters".

Layer 2: Every node in this layer is a fixed node labeled Π, whose output is the product of all incoming signals:

$$0_{2,i} = w_i = \mu_{Ai}(x)\mu_{Bi}(y), i = 1, 2 \tag{4.27}$$

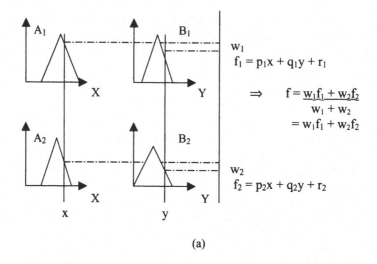

(a)

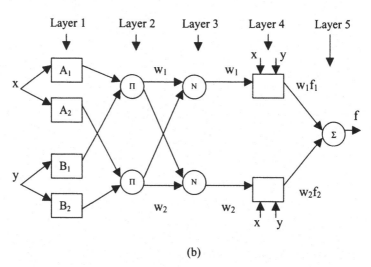

(b)

Fig. 4.21. (a) A two-input Sugeno fuzzy model with 2 rules; (b) equivalent ANFIS architecture (adaptive nodes shown with a square and fixed nodes with a circle)

Each node output represents the firing strength of a fuzzy rule.

Layer 3: Every node in this layer is a fixed node labeled N. The ith node calculates the ratio of the ith rule's firing strength to the sum of all rules' firing strengths:

$$0_{3,i} = w_i = w_i/(w_1 + w_2), i = 1, 2 . \qquad (4.28)$$

For convenience, outputs of this layer are called "normalized firing strengths".

Layer 4: Every node i in this layer is an adaptive node with a node function

$$O_{4,i} = w_i f_i = w_i (p_i x + q_i y + r_i) \,, \qquad (4.29)$$

where w_i is a normalized firing strength from layer 3 and $\{p_i, q_i, r_i\}$ is the parameter set of this node. Parameters in this layer are referred to as "consequent parameters".

Layer 5: The single node in this layer is a fixed node labeled Σ, which computes the overall output as the summation of all incoming signals:

$$\text{overall output} = O_{5,i} = \sum_i w_i f_i = \frac{\sum_i w_i f_i}{\sum_i w_i} \qquad (4.30)$$

Thus we have constructed an adaptive network that is functionally equivalent to a Sugeno fuzzy model. We can note that the structure of this adaptive network is not unique; we can combine layers 3 and 4 to obtain an equivalent network with only four layers. In the extreme case, we can even shrink the whole network into a single adaptive node with the same parameter set. Obviously, the assignment of node functions and the network configuration are arbitrary, as long as each node and each layer perform meaningful and modular functionalities.

The extension from Sugeno ANFIS to Tsukamoto ANFIS is straightforward, as shown in Fig. 4.22, where the output of each rule (f_i, $i = 1$, 2) is induced jointly by a consequent membership function and a firing strength.

4.3.2 Learning Algorithm

From the ANFIS architecture shown in Fig. 4.21(b), we observe that when the values of the premise parameters are fixed, the overall output can be expressed as a linear combination of the consequent parameters. Mathematically, the output f in Fig. 4.21(b) can be written as

$$\begin{aligned}
f &= \frac{w_1 f_1}{w_1 + w_2} + \frac{w_2 f_2}{w_1 + w_2} \\
&= w_1 (p_1 x + q_1 y + r_1) + w_2 (p_2 x + q_2 y + r_2) \\
&= (w_1 x) p_1 + (w_1 y) q_1 + (w_1) r_1 + (w_2 x) p_2 + (w_2 y) q_2 + (w_2) r_2 \quad (4.31)
\end{aligned}$$

which is linear in the consequent parameters p_1, q_1, r_1, p_2, q_2, and r_2. From this observation, we can use a hybrid learning algorithm for parameter estimation in this kind of models (Jang, 1993). More specifically, in the forward pass of the hybrid learning algorithm, node outputs go forward until layer 4 and the consequent parameters are identified by the least-squares method. In the backward pass, the error signals propagate backward and the premise parameters are updated by gradient descent.

It has been shown (Jang, 1993) that the consequent parameters identified in this manner are optimal under the condition that the premise parameters are fixed. Accordingly, the hybrid approach converges much faster since

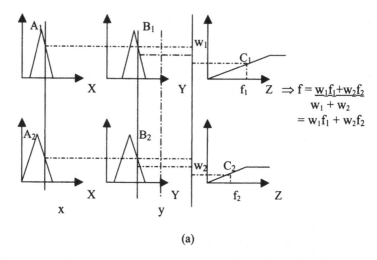

$$\Rightarrow f = \frac{w_1 f_1 + w_2 f_2}{w_1 + w_2}$$
$$= w_1 f_1 + w_2 f_2$$

(a)

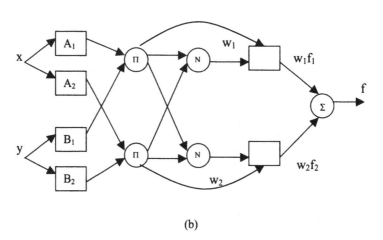

(b)

Fig. 4.22. (a) A two-input Tsukamoto fuzzy model with two rules; (b) equivalent ANFIS architecture

it reduces the search space dimensions of the original pure backpropagation method. For Tsukamoto ANFIS, this can be achieved if the membership function on the consequent part of each rule is replaced by a piecewise linear approximation with two consequent parameters.

As we discussed earlier, under certain minor conditions, an RBFN is functionally equivalent to a fuzzy system, and thus to ANFIS. This functional equivalence provides a shortcut for better understanding both ANFIS and RBFNs in the sense that development in either literature cross-fertilize the other (Jang, Sun & Mizutani, 1997).

Finally, we have to mention that it has been shown that the ANFIS methodology can be viewed as universal approximator (Jang, Sun & Mizutani, 1997). More specifically, it has been shown that when the number of rules is not restricted, a zero-order Sugeno model has unlimited approximation power for matching any non-linear function arbitrarily well on a compact set. This fact is intuitively reasonable. However, the mathematical proof can be made by showing that ANFIS satisfies the well-known Stone-Weierstrass theorem (Kantorovich & Akilov, 1982).

We will now show a simple example to illustrate the ANFIS methodology. We will use as training data the set of points shown in Fig. 4.5 We will use a network of 20 nodes, 4 rules, 4 Gaussian membership functions, and 16 unknown parameters. The complete network is shown in Fig. 4.23, in which we can clearly see all the details mentioned above. We have to mention that the ANFIS methodology is been used here to obtain a first order Sugeno model. In Fig. 4.24, we can appreciate the rate of convergence of ANFIS as the error is plotted against the number of epochs. From this figure it is clear that ANFIS can achieve a comparable error (with the previous methods in this chapter) in only 20 epochs, which is a lot less than the 1000 epochs required by the networks presented before (for the same example). In Fig. 4.25 we can see the final function approximation achieved by the ANFIS method, which is very good. In Fig. 4.26 we show the non-linear surface of the final fuzzy

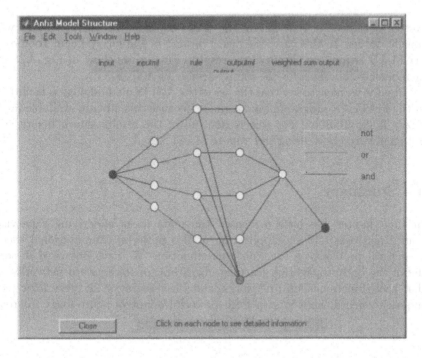

Fig. 4.23. Architecture of network for the ANFIS method

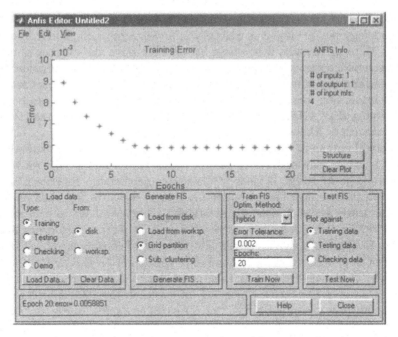

Fig. 4.24. Convergence of ANFIS (final SSE = 0.0058851)

system obtained by ANFIS. In Fig. 4.27, we show the use of ANFIS with specific values. In this case, the "rule viewer" of the Fuzzy Logic Toolbox of MATLAB is used to obtain these results. Finally, we show in Fig. 4.28 the membership functions for the input variable of ANFIS.

Finally, we have to say that the use of the ANFIS methodology is facilitated in the MATLAB programming language because it is already available in the Fuzzy Logic Toolbox. For this reason, all of the results shown before were obtained very easily using this tool of MATLAB.

4.4 Summary

In this chapter, we have presented the main ideas underlying supervised neural networks and the application of this powerful computational theory to general problems in function approximation. We have discussed in some detail the backpropagation learning algorithm for feedforward networks, radial basis function neural networks, and the integration of fuzzy logic techniques to neural networks to form powerful adaptive neuro-fuzzy inference

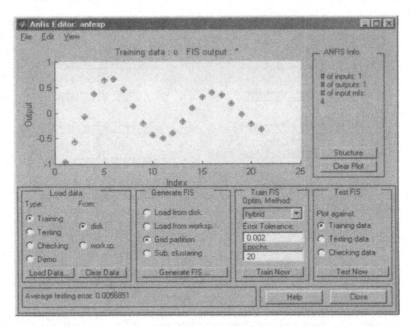

Fig. 4.25. Final function approximation achieved by the ANFIS method

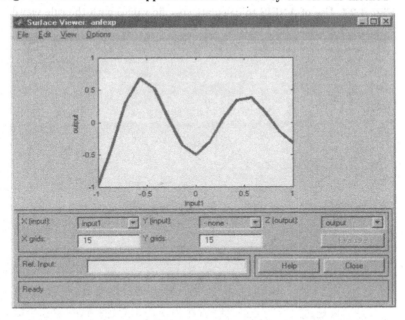

Fig. 4.26. Non-linear surface obtained by the ANFIS method

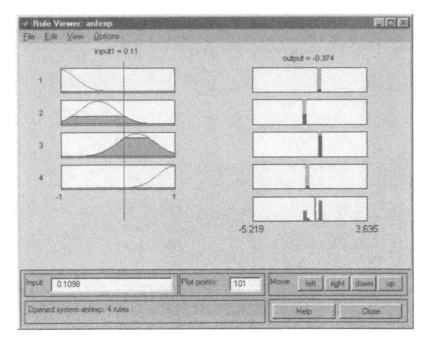

Fig. 4.27. Use of ANFIS to calculate specific values with the rule viewer

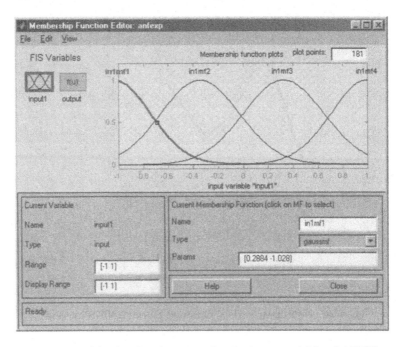

Fig. 4.28. Membership functions for the input variable of ANFIS

systems. In the following chapters, we will show how supervised neural network techniques (in conjunction with other techniques) can be applied to solve real world complex problems in intelligent pattern recognition. This chapter will serve as a basis for the new hybrid intelligent methods that will be described in the chapters at the end of this book.

5

Unsupervised Learning Neural Networks

This chapter introduces the basic concepts and notation of unsupervised learning neural networks. Unsupervised networks are useful for analyzing data without having the desired outputs; in this case, the neural networks evolve to capture density characteristics of a data phase. We will describe in some detail competitive learning networks, Kohonen self-organizing networks, learning vector quantization, and Hopfield networks. We will also show some examples of these networks to illustrate their possible application in solving real-world problems in pattern recognition.

When no external teacher or critic's instruction is available, only input vectors can be used for learning. Such an approach is learning without supervision, or what is commonly referred to as unsupervised learning. An unsupervised learning system evolves to extract features or regularities in presented patterns, without being told what outputs or classes associated with the input patterns are desired. In other words, the learning system detects or categorizes persistent features without any feedback from the environment. Thus unsupervised learning is frequently employed for data clustering, feature extraction, and similarity detection.

Unsupervised learning Neural Networks attempt to learn to respond to different input patterns with different parts of the network. The network is often trained to strengthen firing to respond to frequently occurring patterns, thereby leading to the so-called synonym probability estimators. In this manner, the network develops certain internal representations for encoding input patterns.

5.1 Competitive Learning Networks

With no available information regarding the desired outputs, unsupervised learning networks update weights only on the basis of the input patterns. The competitive learning network is a popular scheme to achieve this type of unsupervised data clustering or classification. Figure 5.1 presents an example

Patricia Melin and Oscar Castillo: *Hybrid Intelligent Systems for Pattern Recognition Using Soft Computing*, StudFuzz **172**, 85–107 (2005)
www.springerlink.com

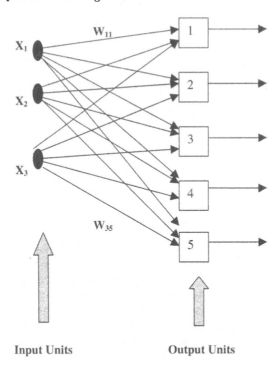

Fig. 5.1. Competitive Learning Network

of competitive learning network. All input units i are connected to all output units j with weight w_{ij}. The number of inputs is the input dimension, while the number of outputs is equal to the number of clusters that the data are to be divided into. A cluster center's position is specified by the weight vector connected to the corresponding output unit. For the simple network in Fig. 5.1, the three-dimensional input data are divided into five clusters, and the cluster centers, denoted as the weights, are updated via the competitive learning rule.

The input vector $X = [x_1, x_2, x_3]^T$ and the weight vector $W_j = [w_{1j}, w_{2j}, w_{3j}]^T$ for an output unit j are generally assumed to be normalized to unit length. The activation value a_j of output unit j is then calculated by the inner product of the input and weight vectors:

$$a_j = \sum_{i=1}^{3} x_i w_{ij} = X^T W_j = W_j^T X \tag{5.1}$$

Next, the output unit with the highest activation must be selected for further processing, which is what is implied by competitive. Assuming that output unit k has the maximal activation, the weights leading to this unit are updated according to the competitive or the so-called winner-take-all learning rule:

$$w_k(t+1) = \frac{w_k(t) + \eta(x(t) - w_k(t))}{||w_k(t) + \eta(x(t) - w_k(t))||} \tag{5.2}$$

The preceding weight update formula includes a normalization operation to ensure that the updated weight is always of unit length. Notably, only the weights at the winner output unit k are updated; all other weights remain unchanged.

The update formula in (5.2) implements a sequential scheme for finding the cluster centers of a data set of which the entries are of unit length. When an input x is presented to the network, the weight vector closest to x rotates toward it. Consequently, weight vectors move toward those areas where most inputs appear and, eventually, the weight vectors become the cluster centers for the data set. In Fig. 5.2 we show this dynamic process.

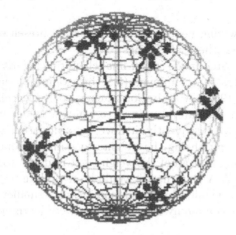

Fig. 5.2. Competitive learning with unit-length vectors. The *dots* represent the input vectors and the *crosses* denote the weight vectors for the five output units in Fig. 5.1. As the learning continues, the five weight vectors rotate toward the centers of the five input clusters

Using the Euclidean distance as a dissimilarity measure is a more general scheme of competitive learning, in which the activation of output unit j is

$$a_j = \left(\sum_{i=1}^{3} (x_i - w_{ij})^2 \right)^{0.5} = \|x - w_j\| \tag{5.3}$$

The weights of the output unit with the smallest activation are updated according to

$$w_k(t+1) = w_k(t) + \eta(x(t) - w_k(t)) \tag{5.4}$$

In the (5.4), the wining unit's weights shift toward the input x. In this case, neither the data nor the weights must be of unit length.

A competitive learning network performs an on-line clustering process on the input patterns. When the process is complete, the input data are divided into disjoint clusters such that similarities between individuals in the same cluster are larger than those in different clusters. Here two metrics of similarity are introduced: the similarity measure of inner product in (5.1) and the dissimilarity measure of the Euclidean distance in (5.3). Obviously, other metrics can be used instead, and different selections lead to different clustering results. When the Euclidean distance is adopted, it can be proved that the update formula in (5.4) is actually an on-line version of gradient descent that minimizes the following objection function:

$$E = \sum_p \|w_{f(x_p)} - x_p\|^2 \tag{5.5}$$

where $f(x_p)$ is the wining neuron when input x_p is presented and $w_{f(x_p)}$ is the center of the class where xp belongs to.

A large family of batch-mode (or off-line) clustering algorithms can be used to find cluster centers that minimize (5.5) by example K-means clustering algorithm (Ruspini, 1982). A limitation of competitive learning is that some of the weight vectors that are initialized to random values may be far from any input vector and, subsequently, it never gets updated. Such a situation can be prevented by initializing the weights to samples from the input data itself, thereby ensuring that all of the weights get updated when all the input patterns are presented. An alternative would be to update the weights of both the winning and losing units, but use a significantly smaller learning rate h for the losers; this is commonly referred to as leaky learning (Rumelhart & Zipser, 1986).

Dynamically changing the learning rate η in the weight update formula of (5.2) or (5.4) is generally desired. An initial large value of η explores the data space widely; later on, a progressively smaller value refines the weights. The operation is similar to the cooling schedule of simulated annealing, as introduced in Chap. 6. Therefore, one of the following formulas for η is commonly used:

$$\begin{cases} \eta(t) = \eta_o e^{-\alpha t}, & \text{with} \quad \alpha > 0, \text{or} \\ \eta(t) = \eta_o t^{-\alpha}, & \text{with} \quad \alpha \le 0, \text{or} \\ \eta(t) = \eta_o(1 - \alpha t), & \text{with} \quad 0 < \alpha < (\max\{t\})^{-1} \end{cases}$$

Competitive learning lacks the capability to add new clusters when deemed necessary. Moreover, if the learning rate η is a constant, competitive learning does not guarantee stability in forming clusters; the winning unit that responds to a particular pattern may continue changing during training. On the other hand, η, if decreasing with time, may become too small to update cluster centers when new data of a different probability nature are presented. Carpenter and Grossberg referred to such an occurrence as the stability-plasticity dilemma, which is common in designing intelligent learning systems (Carpenter & Grossberg, 1998). In general, a learning system should be plastic, or adaptive in reacting to changing environments; meanwhile, it should be stable to preserve knowledge acquired previously.

If the output units of a competitive learning network are arranged in a geometric manner (such as in a one-dimensional vector or two-dimensional array), then we can update the weights of the winners as well as the neighboring losers. Such a capability corresponds to the notion of Kohonen feature maps, as discussed in the next section.

We illustrate the use of competitive learning networks with a simple example. In Fig. 5.3 we show the initial data randomly distributed within the rectangle $[-0.2, 1.2] \times [0, 1]$. We can also see at the center the initial weights of the network (the crosses indicate the weights). We did specified a-priori the number of clusters as eight. After training the competitive network for 500 epochs we obtained the result shown in Fig. 5.4. In this figure we can see how the weights are arranged in their final positions, which are the centers of the clusters.

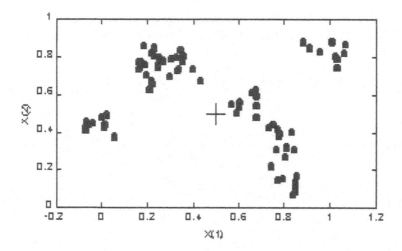

Fig. 5.3. Initial randomly distributed data and weights for the competitive network

5.2 Kohonen Self-Organizing Networks

Kohonen self-organizing networks, also known as Kohonen feature maps or topology-preserving maps, are another competition-based network paradigm for data clustering (Kohonen, 1982, 1984). Networks of this type impose a neighborhood constraint on the output units, such that a certain topological property in the input data is reflected in the output units' weights.

In Fig. 5.5 we show a relatively simple Kohonen self-organizing network with 2 inputs and 49 outputs. The learning procedure of Kohonen feature maps is similar to that of competitive learning networks. That is, a similarity

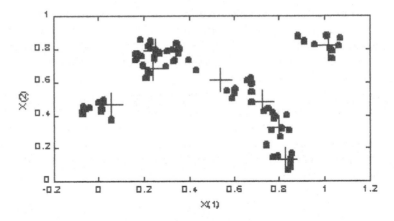

Fig. 5.4. Final distribution of the 8 clusters after training for 500 epochs

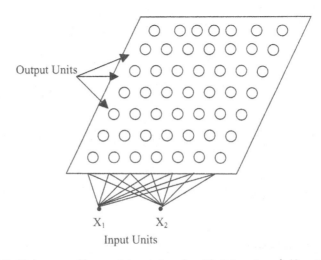

Fig. 5.5. Kohonen self-organizing network with 2 input and 49 output units

(dissimilarity) measure is selected and the winning unit is considered to be the one with the largest (smallest) activation.

For Kohonen feature maps, however, we update not only the winning unit's weights but also all of the weights in a neighborhood around the wining units. The neighborhood's size generally decreases slowly with each iteration, as indicated in Fig. 5.6.

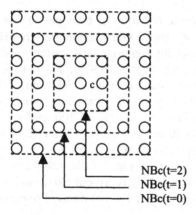

Fig. 5.6. In Kohonen self-organizing network the size of a neighborhood around a winning unit decreases gradually with each iteration

A sequential description of how to train a Kohonen self-organizing network is as follows:

Step 1: Select the wining output unit as the one with the largest similarity measure (or smallest dissimilarity measure) between all weight vectors w_i and the input vector x. If the Euclidean distance is chosen as the dissimilarity measure, then the winning unit c satisfies the following equation:

$$\|x - w_c\| = \min_i \|x - w_i\|$$

where the index c refers to the winning unit.

Step 2: Let $N\,B_c$ denote a set of index corresponding to a neighborhood around winner c. The weights of the winner and its neighboring units are then updated by

$$\Delta wi = \eta(x - w_i)\,, \quad i \in NB_c$$

where η is a small positive learning rate. Instead of defining the neighborhood of a winning unit, we can use a neighborhood function $\Omega c(i)$ around a winning unit c. For instance, the Gaussian function can be used as the neighborhood function:

$$\Omega c(i) = \exp\left(\frac{-\|p_i - p_c\|^2}{2\sigma^2}\right)$$

where p_i and p_c are the positions of the output units i and c, respectively, and σ reflects the scope of the neighborhood. By using the neighborhood function, the update formula can be rewritten as

$$\Delta wi = \eta\Omega_c(i)(x - w_i)\,,$$

where i is the index for all output units.

The most well-known application of Kohonen self-organizing networks is Kohonen's attempt to construct a neural phonetic typewriter (Kohonen, 1988) that is capable of transcribing speech into written text from an unlimited vocabulary, with an accuracy of 92% to 97%. The network has also been used to learn ballistic arm movements (Ritter & Schulten, 1987).

We first illustrate the use of self-organizing networks with a very simple example. We have as initial data a set of 100 uniformly distributed points in the range $[0, \pi/2]$, as shown in Fig. 5.7. After training a one-dimensional Kohonen self-organizing map of 10 neurons for 1000 epochs, we arrived to the final configuration shown in Fig. 5.8. In this figure we can see the final distribution of the neurons, which represent the centers of the clusters. From this last figure we can notice that the points were organized around 10 evenly distributed clusters.

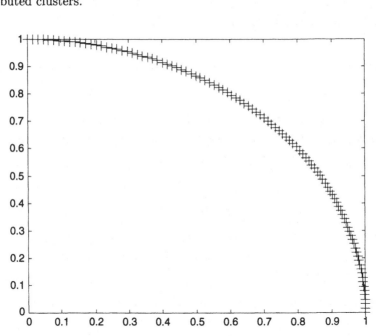

Fig. 5.7. Initial data set of 100 uniformly distributed points in the range $[0, \pi/2]$

Our second example illustrates the application of a bi-dimensional Kohonen self-organizing map. We have as initial training data a set of 1000 randomly distributed points in the square $[-1, 1] \times [-1, 1]$, this initial situation is shown in Fig. 5.9. After applying a 5×6 bi-dimensional self-organizing map for 1000 epochs, we arrived to the final configuration shown in Fig. 5.10. In this last figure we can see very clearly the distribution of the 30 centers of the clusters after training. In this case, the final configuration of the clusters is not uniformly distributed due to the characteristics of the initial distribution of points.

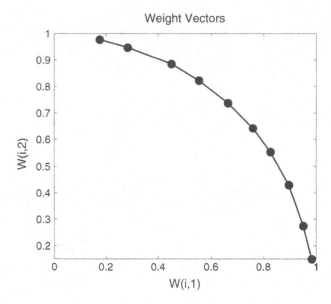

Fig. 5.8. After training a self-organizing map of 10 neurons for 1000 epochs

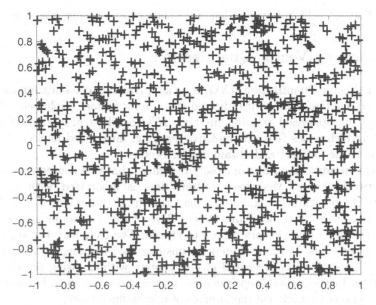

Fig. 5.9. Initial set of 1000 distributed points in the square $[-1, 1] \times [-1, 1]$

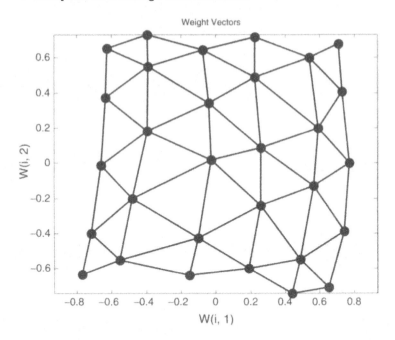

Fig. 5.10. Final configuration of the clusters after training the map

5.3 Learning Vector Quantization

Learning Vector Quantization (LVQ) is an adaptive data classification method based on training data with desired class information (Kohonen, 1989). Although a supervised training method, LVQ employs unsupervised data-clustering techniques (e.g., competitive learning, introduced in Sect. 5.1) to preprocess the data set and obtain cluster centers.

LVQ's network architecture closely resembles that of a competitive learning network, except that each output unit is associated with a class. Figure 5.11 presents an example, where the input dimension is 2 and the input space is divided into four clusters. Two of the clusters belong to class 1, and the other two clusters belong to class 2. The LVQ learning algorithm involves two steps. In the first step, an unsupervised learning data clustering method is used to locate several cluster centers without using the class information. In the second step, the class information is used to fine-tune the cluster centers to minimize the number of misclassified cases.

During the first step of unsupervised learning, any of the data clustering techniques introduced in this chapter can be used to identify the cluster centers (or weight vectors leading to output units) to represent the data set with no class information. The numbers of clusters can either be specified a priori or determined via a cluster technique capable of adaptively adding new clusters when necessary. Once the clusters are obtained, their classes must be labeled

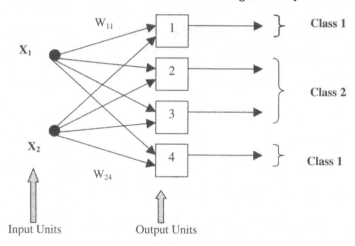

Fig. 5.11. LVQ network representation

before moving to the second step of supervised learning. Such labeling is achieved by the so-called "voting method" (i.e., a cluster is labeled class k if it has data points belonging to class k as a majority within the cluster). The clustering process for LVQ is based on the general assumption that similar input patterns generally belong to the same class.

During the second step of supervised learning, the cluster centers are fine-tuned to approximate the desired decision surface. The learning method is straightforward. First, the weight vector (or cluster center) $\mathbf{w}$ that is closest to the input vector $\mathbf{x}$ must be found. If $\mathbf{x}$ and $\mathbf{w}$ belong to the same class, we move $\mathbf{w}$ toward $\mathbf{x}$; otherwise we move $\mathbf{w}$ away from the input vector $\mathbf{x}$.

After learning, an LVQ network classifies an input vector by assigning it to the same class as the output unit that has the weight vector (cluster center) closest to the input vector. Figure 5.12 illustrates a possible distribution of data set and weights after training.

A sequential description of the LVQ method is the following:

Step 1: Initialize the cluster centers by a clustering method.
Step 2: Label each cluster by the voting method.
Step 3: Randomly select a training input vector $\mathbf{x}$ and find k such that $\|\mathbf{x} - \mathbf{w}_k\|$ is a minimum.
Step 4: If $\mathbf{x}$ and $\mathbf{w}_k$ belong to the same class, update $\mathbf{w}_k$ by

$$\Delta \mathbf{w}_k = \eta(x - \mathbf{w}_k) \,.$$

Otherwise, update $\mathbf{w}_k$ by

$$\Delta \mathbf{w}_k = -\eta(x - \mathbf{w}_k) \,.$$

The learning rate η is a positive small constant and should decrease in value with each respective iteration.

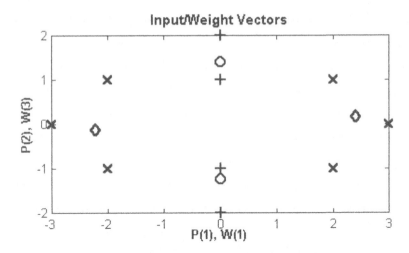

Fig. 5.12. Data distribution and classification obtained by LVQ

Step 5: If the maximum number of iterations is reached, stop. Otherwise, return to step 3.

Two improved versions of LVQ are available; both of them attempt to use the training data more efficiently by updating the winner and the runner-up (the next closest vector) under a certain condition. The improved versions are called LVQ2 and LVQ3 (Kohonen 1990).

We will consider a simple example to illustrate the application of the LVQ network for classification purposes. Lets assume we have the data shown in Fig. 5.13, which represents the 5 letters: O, C, A, L and I. We will apply a 5 neuron LVQ network to classify these letters. We show in Fig. 5.14 the initial step of the LVQ training in which there is only one cluster at the center. In Fig. 5.15 we show the final result after training the LVQ network. In this figure it is clearly seen that we have now five clusters, one for each of the letters. Also, the LVQ network has classified the letters in two types, which are the vowels (A, I and O) and the consonants (C and L). The vowels are indicated with circles and the consonants with diamonds.

We will now consider a more complicated and realistic example of the application of LVQ neural networks. The San Quintin valley is a region of Baja California, Mexico in which tomato is the main export product. The tomato is taken from the fields to the food processing plants, in which according to their quality a classification is performed. Tomato is classified in to four categories: export quality, national quality, regional, and salsa quality. This classification is very important because the net profit from tomato production depends on the quality evaluation been done appropriately.

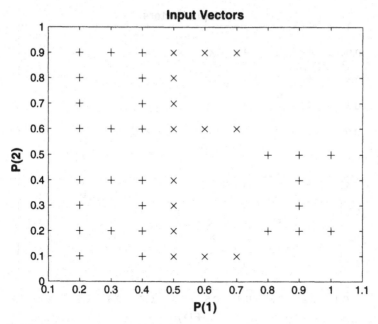

Fig. 5.13. Data set representing 5 letters from the alphabet: O, C, A, L, and I

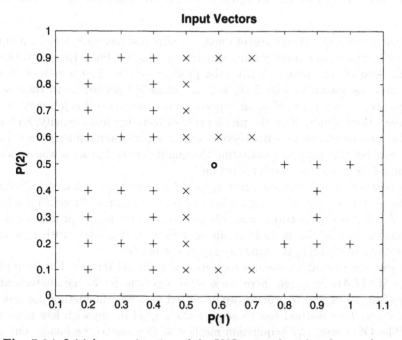

Fig. 5.14. Initial approximation of the LVQ network with only one cluster

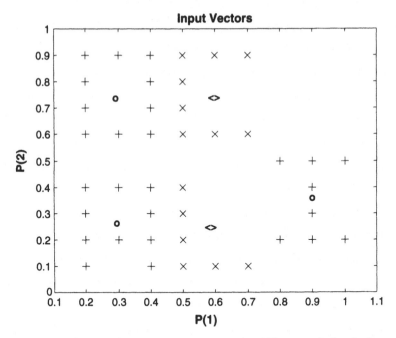

Fig. 5.15. Final classification achieved with the LVQ network for the letters

Traditionally quality control of tomato production has been done by manually selecting the good quality tomatoes from production lines. Human workers visually classify the tomatoes from the production line. The tomatoes move on a belt, as shown in Fig. 5.16, and are classified according to their size and to their appearance. Human experts know how to do this job very well. However, they usually have the problems of objectivity, homogeneity, and accuracy, mentioned before. Also, there is a need of performing quality control in less time. For this reason, automating the quality control process was decided to reduce costs and times of production.

A window of opportunity was recognized for the application of soft computing techniques for achieving automated quality control of tomato production. A LVQ neural network was selected to perform image processing and classification. Also, fuzzy logic techniques were used to deal with inherent uncertainty in deciding the final quality of the product.

There are several techniques for processing digital images. For example, in the MATLAB language there are several functions for image analysis and processing. To begin the process, we first need to define the method for image acquisition. This method has to acquire the digital image with less noise as possible. Of course, any acquisition method adds noise to the image, but we need to use a method that minimizes this noise.

For a real time system, like for tomato classification, image acquisition has to be done with a digital camera, but the scene has to be reflection free

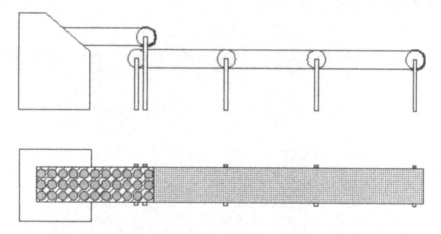

Fig. 5.16. Belt of a tomato production line

and with uniform illumination. We show in Fig. 5.17(a) the image of a tomato acquired with a digital camera. Figure 5.17(b) shows the same image, but with intensity values adjusted for better processing. After the image is acquired, then we perform processing operations to calculate the area of the image and estimate the size of the tomato.

Now we describe how the LVQ network was applied to achieve image classification. First, we did make a choice about which pixels of the image to use. For tomato image we used ten pixels as input to the network. We show in Fig. 5.18(a) the tomato image with the pixels used as input. Of course, the final choice of pixels was achieved after experimentation, and comparison of results. We consider that the final choice of pixels is representative of the image, or at least this was true for the purpose of this work. We also show in Fig. 5.18(b) the final training of the LVQ network after 1000 cycles of learning.

The LVQ neural network basically classifies the tomato images according to their similarities. However, there exists uncertainty in this classification. For this reason, the use of fuzzy logic is justified to complement the LVQ network.

We formulated a set of fuzzy if then rules to automate quality evaluation of the tomato according to the results of the LVQ network. We used as input linguistic variables the area of the image (size of the tomato), maturity of the tomato, and defects of the tomato. We used as output linguistic variable the class of the tomato. We used a Mamdani type fuzzy system with seven fuzzy rules, which were formulated to model experts in this task. We show in Fig. 5.19 the general architecture of the fuzzy system for automated quality control. In this figure it is clearly shown the use of the three inputs and one output.

Now we describe the choice of the linguistic terms and membership functions for each of the linguistic variables in the fuzzy system. First, for the area

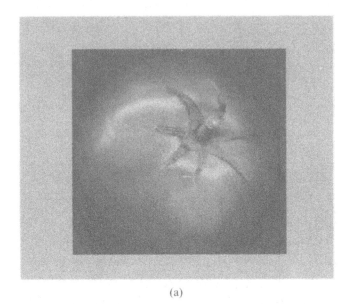

(a)

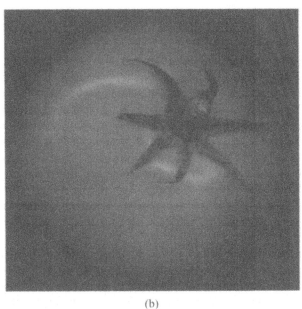

(b)

Fig. 5.17. (a) Image acquired by digital camera, (b) Image with intensity values adjusted

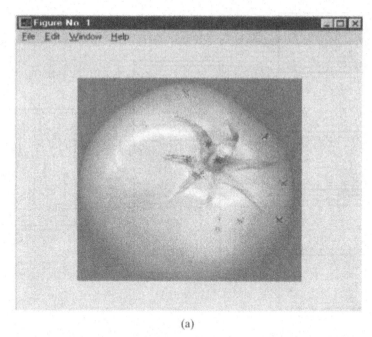

(a)

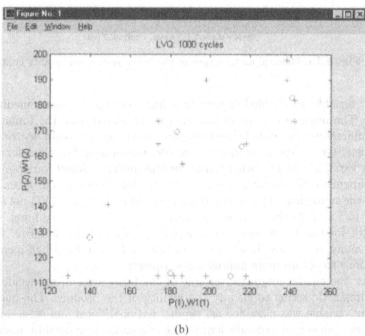

(b)

Fig. 5.18. (a) Image with the choice of pixels, (b) Final training of the LVQ network

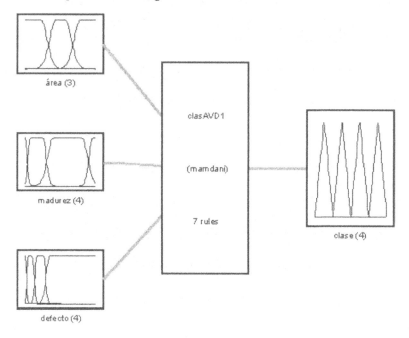

System olasAVD 1: 3 inputs, 1 outputs, 7 rules

Fig. 5.19. General architecture of the fuzzy system for quality control

input variable we decided to use three linguistic terms: small, medium, and large. The numerical value of this area is calculated from the tomato image (in millimeters) and this is used to calculate the membership degree to these linguistic terms. We show in Fig. 5.20 the membership functions for the area input variable. On the other hand, for the maturity input variable we used four linguistic terms: very mature, mature, less mature, and, green. In this case, the numerical value of maturity is obtained from the LVQ network. We show in Fig. 5.20 the membership functions of the maturity input variable. Finally, for the defect input variable we used four linguistic terms: few, regular, many, too many. In all the cases, we used generalized bell membership functions to obtain more approximation power.

For the output linguistic variable of the fuzzy system, we decided to use four linguistic terms to classify the quality of the product. The quality linguistic variable was classified as: export, national, local and salsa quality. In this case, we considered sufficient to use triangular membership functions for the output variable because we only needed to do the classification. We show in Fig. 5.21 the membership functions for the quality linguistic variable.

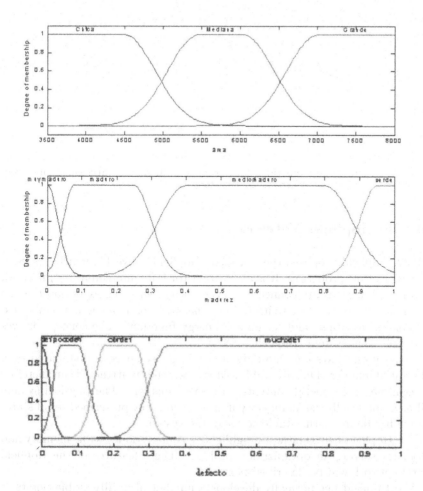

Fig. 5.20. Membership functions for the input linguistic variables of the fuzzy system

The new approach has been tested with the specific case of automating the quality control of tomato in a food processing plant with excellent results. The accuracy of the classification has been improved, using the hybrid approach, from 85% to approximately 95%. On the other hand, the time required to perform this classification process has been reduced more than 15% with the automation of the process. Finally, we have to say that this hybrid approach can be used for similar problems in quality control.

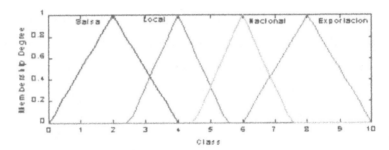

Fig. 5.21. Membership functions for the output linguistic variable of the fuzzy system

5.4 The Hopfield Network

In 1982, Hopfield proposed the so-called "Hopfield network", which possesses *auto-associative* properties. It is a recurrent (or fully interconnected) network in which all neurons are connected to each other, with the exception that no neuron has any connection to itself. In the network configuration, he embodied the *physical principle*, and set up an "energy function". The concept derives from a physical system (Hopfield, 1982):

"Any physical system whose dynamics in phase space is dominated by a substantial number of locally stable states to which it is attracted can therefore be regarded as a general content-addressable memory. The physical system will be a potentially useful memory if, in addition, any prescribed set of states can readily be made the stable states of the system."

The Hopfield neural network highlights a content-addressable memory and a tool for solving an optimization problem. These features of the Hopfield neural network will be described next.

The Hopfield net normally develops a number of locally stable points in state space. Because the dynamics of the network minimize "energy", other points in state space drain into the stable points (called "attractors" or "wells"), which are (possibly local) energy minima.

The Hopfield net realizes the operation of a "content-addressable" (auto-associative) memory in the sense that newly presented input patterns (or arbitrary initial states) can be connected to the appropriate patterns stored in memories (i.e., attractors or stable states). The presented input pattern vector cannot escape from a region, what we call a "basin of attraction", configured by each attractor (Fig. 5.22). In other words, the neural network produces a desired memory pattern in response to a given pattern. The initial network state is assumed to lie within the reach of the fixed attractors' basins of attraction. That is, the entire configuration space is divided into different basins of attraction. Notably, an attractor is regarded as a fixed point if it is unique in state space. In general, however, an attractor may be "chaotic' (called a strange attractor) or may be a "limit cycle" of a periodic sequence of

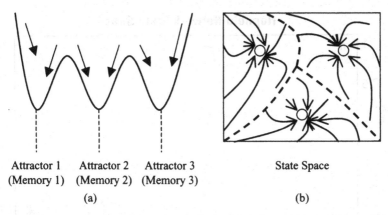

Attractor 1	Attractor 2	Attractor 3	State Space
(Memory 1)	(Memory 2)	(Memory 3)	
	(a)		(b)

Fig. 5.22. Images of energy terrains configured by three attractors in a Hopfield network. The *circles* denote stable states where patterns are memorized. The network state moves in the direction of the arrows to one of the attractors, which is determined by the starting point (i.e., the given input pattern)

states. In terms of storage capacity, the number of memories is estimated to be nearly 10% to 20 % of the number of neurons in the Hopfield net (Denker, 1986).

We will illustrate the basic ideas of Hopfield networks with two simple examples. Lets consider first the case of a simple two-neuron Hopfield net that has two stable states. We show in Fig. 5.23 the representation in state space of the two neurons, i.e., the attractors or wells of the dynamical system. We also show in Fig. 5.24 the simulation of the two-neuron Hopfield neural network, which clearly shows the two basins of attraction. We can interpret this figure as follows: initial points which correspond to one of the basins of attraction are pulled to that neuron (the arrows indicate the directions).

The example illustrates the very simple case of a two-neuron Hopfield network. Of course, increasingly complicated networks with more than two neurons may develop many stable states (attractors) to memorize several patterns.

Lets consider as a second example three-neuron Hopfield network in three-dimensional space. The network has two stable states: the points $(-1, 1, -1)$ and $(1, 1, -1)$, which will divide the state space in two equal basins of attraction. We show in Fig. 5.25 the state space for this Hopfield network. In Fig. 5.26 we show how the initial points are attracted to one of these points. The directions are indicated with arrows. Actually, there is another stable point, which is $(0, 1, -1)$, but the set of points, forming its basin of attraction, has measure zero.

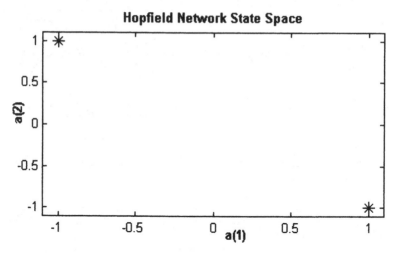

Fig. 5.23. Hopfield network state space

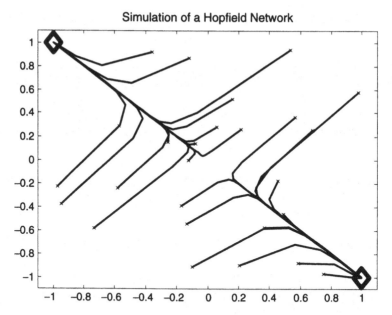

Fig. 5.24. Simulation of a two-neuron Hopfield Network

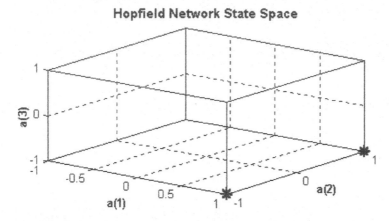

Fig. 5.25. Three-dimensional state space for the Hopfield network

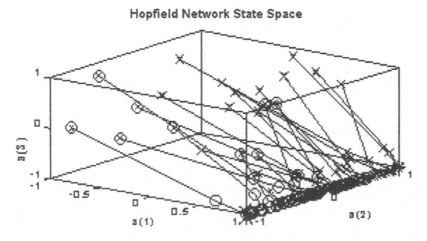

Fig. 5.26. Simulation of the Hopfield network for different initial conditions

5.5 Summary

We have presented in this chapter the basic concepts and theory of unsupervised learning neural networks. We have described in some detail competitive learning networks, Kohonen self-organizing networks, learning vector quantization networks, and Hopfield neural networks. Unsupervised learning is useful for analyzing data without having desired outputs; the neural networks evolve to capture density characteristics of a data set. We have shown some examples of how to use this type of unsupervised networks. We will describe in future chapters some applications of unsupervised neural networks to real world problems related to pattern recognition.

6

Modular Neural Networks

We describe in this chapter the basic concepts, theory and algorithms of modular and ensemble neural networks. We will also give particular attention to the problem of response integration, which is very important because response integration is responsible for combining all the outputs of the modules. Basically, a modular or ensemble neural network uses several monolithic neural networks to solve a specific problem. The basic idea is that combining the results of several simple neural networks we will achieve a better overall result in terms of accuracy and also learning can be done faster. For pattern recognition problems, which have great complexity and are defined over high dimensional spaces, modular neural networks are a great alternative for achieving the level of accuracy and efficiency needed for real-time applications. This chapter will serve as a basis for the modular architectures that will be proposed in later chapters for specific pattern recognition problems.

6.1 Introduction

The results of the different applications involving Modular Neural Networks (MNN) lead to the general evidence that the use of modular neural networks implies a significant learning improvement comparatively to a single NN and especially to the backpropagation NN. Indeed, to constrain the network topology on connectivity, increases the learning capacity of NN and allow us to apply them to large-scale problems (Happel & Murre, 1994). This is highly confirmed by the experience carried out by (Barna & Kaski, 1990), which shows that a random pruning of connections before any learning improves significantly the network's performance. Also, we can note that (Feldman, 1989) and (Simon, 1981) argue that a complex behavior requires bringing together several different kinds of knowledge and processing, which is, of course, not possible without structure (modularity).

Usually the MNN implementations are based on the "divide and conquer" principle, which is well known in computer science. This principle consists

Patricia Melin and Oscar Castillo: *Hybrid Intelligent Systems for Pattern Recognition Using Soft Computing*, StudFuzz **172**, 109–129 (2005)
www.springerlink.com

first in breaking down a task into smaller and less complex subtasks, to make learn each task by different experts (i.e. NN modules) and then, to reuse the learning of each subtask to solve the whole problem. For example, (Jenkins & Yuhas, 1993) has shown that NN training can be greatly simplified by identifying subtasks in the problem and embedding them into the network structure.

This method can only be applied if the a-priori knowledge concerning the task is sufficiently precise to enable a split up of the task into subtasks. If one can separate a task in distinct subtasks, each task can be trained off-line and later integrate in the global architecture (). This enables an acceleration of the learning (Fogelman-Soulie, 1993).

There exists a lot of neural network architectures in the literature that work well when the number of inputs is relatively small, but when the complexity of the problem grows or the number of inputs increases, their performance decreases very quickly. For this reason, there has also been research work in compensating in some way the problems in learning of a single neural network over high dimensional spaces.

In the work of (Sharkey, 1999), the use of multiple neural systems (Multi-Nets) is described. It is claimed that multi-nets have better performance or even solve problems that monolithic neural networks are not able to solve. It is also claimed that multi-nets or modular systems have also the advantage of being easier to understand or modify, if necessary.

In the literature there is also mention of the terms "ensemble" and "modular" for this type of neural network. The term "ensemble" is used when a redundant set of neural networks is utilized, as described in (Hansen & Salamon, 1990). In this case, each of the neural networks is redundant because it is providing a solution for the same task, as it is shown in Fig. 6.1.

On the other hand, in the modular approach, one task or problem is decompose in subtasks, and the complete solution requires the contribution of all the modules, as it is shown in Fig. 6.2.

In this section we describe a particular class of "modular neural networks", which have a hierarchical organization comprising multiple neural networks; the architecture basically consists of two principal components: local experts and an integration unit, as illustrated in Fig. 6.3. A variety of modular connectionist architectures have been discussed, and thus such diverse names as "committees of networks", "adaptive mixtures", and "hierarchical mixtures of experts" have all been mentioned.

In general, the basic concept resides in the idea that combined (or averaged) estimators may be able to exceed the limitation of a single estimator. The idea also shares conceptual links with the "divide and conquer" methodology. Divide-and-conquer algorithms attack a complex problem by dividing it into simpler problems whose solutions can be combined to yield a solution to the complex problem. In other words, the central idea is task decomposition. When using a modular network, a given task is split up among several local experts NNs. The average load on each NN is reduced in compar-

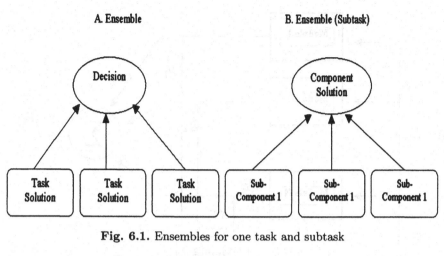

Fig. 6.1. Ensembles for one task and subtask

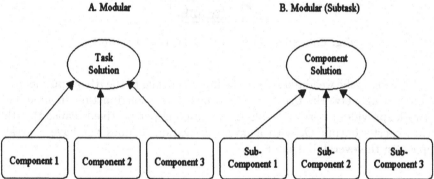

Fig. 6.2. Modular approach for task and subtask

ison with a single NN that must learn the entire original task, and thus the combined model may be able to surpass the limitation of a single NN. The outputs of a certain number of local experts (O_i) are mediated by an integration unit. The integrating unit puts those outputs together using estimated combination weights (g_i). The overall output Y of the modular network is given by

$$Y_i = \Sigma g_i O_I \tag{6.1}$$

In the work by (Nowlan, Jacobs, Hinton, & Jordan, 1991) it is described modular networks from a competitive mixture perspective. That is, in the gating network, they used the "softmax" activation function, which was introduced by (McCullagh & Nelder, 1994). More precisely, the gating network uses a softmax activation g_i of the ith output unit given by

$$G_i = \exp(ku_i)/\Sigma_j \exp(ku_j) \tag{6.2}$$

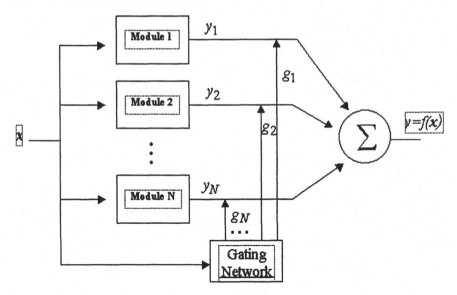

Fig. 6.3. Architecture of modular neural network

Where u_i is the weighted sum of the inputs flowing to the ith output neuron of the gating network. Use of the softmax activation function in modular networks provides a sort of "competitive" mixing perspective because the ith local expert's output O_i with a minor activation u_i does not have a great impact on the overall output Y_i.

6.2 Overview of Multiple Neural Networks

We give in this section a brief overview of the research work that has been done in the past few years in the area of multiple neural networks. We will describe the main architectures, the methods for response integration, the communication between modules and other important concepts about MNN.

6.2.1 Multiple Neural Networks

In this approach we can find networks that use strongly separated architectures. Each neural network works independently in its own domain. Each of the neural networks is build and trained for a specific task. The final decision is based on the results of the individual networks, called agents or experts.

One example of this decision method is described by (Schmidt, 1996), as shown in Fig. 6.4, where a multiple architecture is used. One module consists of a neural network trained for recognizing a person by the voice, while the other module is a neural network trained for recognizing a person by the image.

Network Expert 2

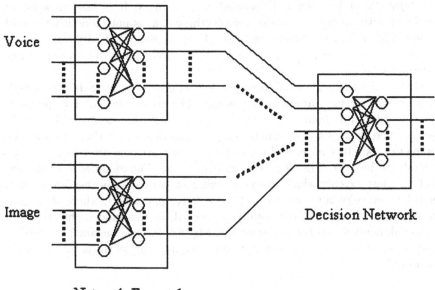

Voice

Image

Decision Network

Network Expert 1

Fig. 6.4. Multiple networks for voice and image

The outputs by the experts are the inputs to the decision network, which is the one making the decision based on the outputs of the expert networks. In this case, the decision neural network has to be trained in such a way as to be able to combine correctly the outputs of the expert networks.

6.2.2 Main Architectures with Multiple Networks

Within multiple neural networks we can find three main classes of this type of networks (Fu & Lee, 2001):

- Mixture of Experts (ME): The mixture of experts can be viewed as a modular version of the multi-layer networks with supervised training or the associative version of competitive learning. In this design, the local experts are trained with the data sets to mitigate weight interference from one expert to the other.
- Gate of Experts: In this case, an optimization algorithm is used for the gating network, to combine the outputs from the experts.
- Hierarchical Mixture of Experts: In this architecture, the individual outputs from the experts are combined with several gating networks in a hierarchical way.

6.2.3 Modular Neural Networks

The term "Modular Neural Networks" is very fuzzy. It is used in a lot of ways and with different structures. Everything that is not monolithic is said to be modular. In the research work by (Boers & Kuiper, 1992), the concept of a modular architecture is introduced as the development of a large network using modules.

One of the main ideas of this approach is presented in (Schmidt, 1996), where all the modules are neural networks. The architecture of a single module is simpler and smaller than the one of a monolithic network. The tasks are modified in such a way that training a subtask is easier than training the complete task. Once all modules are trained, they are connected in a network of modules, instead of using a network of neurons. The modules are independent to some extent, which allows working in parallel. Another idea about modular networks is presented by (Boers & Kuiper, 1992), where they used an approach of networks not totally connected. In this model, the structure is more difficult to analyze, as shown in Fig. 6.5. A clear separation between modules can't be made. Each module is viewed as a part of the network totally connected.

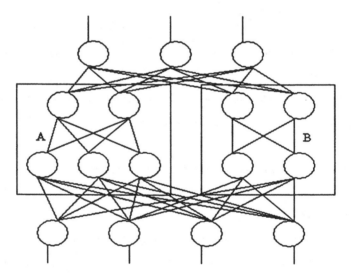

Fig. 6.5. One particular type of modular network

In this figure, we can appreciate two different sections from the monolithic neural network, namely A and B. Since there are no connections between both parts of the network, the dimensionality (number of weights) is reduced. As a consequence the required computations are decreased and speed of convergence is increased.

6.2.4 Advantages of Modular Neural Networks

A list of advantages of modular networks is given below:

- They give a significant improvement in the learning capabilities, over monolithic neural networks, due to the constraints imposed on the modular topology.
- They allow complex behavior modeling, by using different types of knowledge, which is not possible without using modularity.
- Modularity may imply reduction of number of parameters, which will allow and increase in computing speed and better generalization capabilities.
- They avoid the interference that affects "global" neural networks.
- They help determine the activity that is being done in each part of the system, helping to understand the role that each network plays within the complete system.
- If there are changes in the environment, modular networks enable changes in an easier way, since there is no need to modify the whole system, only the modules that are affected by this change.

6.2.5 Elements of Modular Neural Networks

When considering modular networks to solve a problem, one has to take into account the following points (Ronco & Gawthrop, 1995):

- Decompose the main problem into subtasks.
- Organizing the modular architecture, taking into account the nature of each subtask.
- Communication between modules is important, not only in the input of the system but also in the response integration.

In the particular case of this paper, we will concentrate in more detail in the third point, the communication between modules, more specifically information fusion at the integrating module to generate the output of the complete modular system.

6.2.6 Main Task Decomposition into Subtasks

Task Decomposition can be performed in three different ways, as mentioned by (Lu & Ito, 1998):

- *Explicit Decomposition*: In this case, decomposition is made before learning and requires that the designer has deep knowledge about the problem. Of course, this maybe a limitation if there isn't sufficient knowledge about the problem.
- *Automatic Decomposition*: In this case, decomposition is made as learning is progressing.

- *Decomposition into Classes*: This type of decomposition is made before learning, a problem is divided into a set of sub-problems according to the intrinsic relations between the training data. This method only requires knowledge about the relations between classes.

6.2.7 Communication Between Modules

In the research studies made by (Ronco & Gawthrop, 1995), several ways of achieving communication between modules are proposed. We can summarize their work by mentioning the following critical points:

1. How to divide information, during the training phase, between the different modules of the system.
2. How to integrate the different outputs given by the different modules of the system to generate the final output of the complete system.

6.2.8 Response Integration

Response integration has been considered in several ways, as described by (Smith & Johansen, 1997) and we can give the following list:

- Using Kohonen's self organizing maps, Gaussian mixtures, etc.
- The method of "Winner Takes All", for problems that require similar tasks.
- Models in series, the output of one module is the input to the following one.
- Voting methods, for example the use of the "Softmax" function.
- Linear combination of output results.
- Using discrete logic.
- Using finite state automata.
- Using statistical methods.
- Using fuzzy logic.

We will concentrate more in this chapter on the use of fuzzy logic as a method for achieving response integration. The idea of considering fuzzy logic is that we want to take into account the uncertainty of the results of the neural network modules, as well as the intrinsic uncertainty of pattern recognition problems. However, we will review briefly some of the architectures that we have mentioned before, so that the reader can get a general idea of what is available in the literature of MNN.

We show in Fig. 6.6 an architecture in which we use a series neural networks. This architecture is called "neural network model in series" and the basic idea of this architecture is that we can use the outputs of some neural networks to be the inputs of other networks and son on. This architecture was initially proposed by (Schmidt, 1996) and there has been some research on applying it to real problems, but there are only very few publications in which there are reports of the use of this architecture.

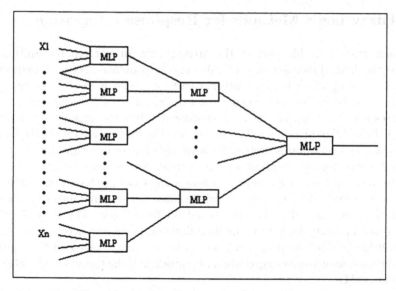

Fig. 6.6. Model of a series of neural networks

Another approach is to use discrete logic. In this case, a logic function is used to make the selection of which module is giving the output of the system. This method has problems when there are discontinuities in the transitions between the use of the different modules. One possible scheme is the one using a multiplexer, like the one in Fig. 6.7.

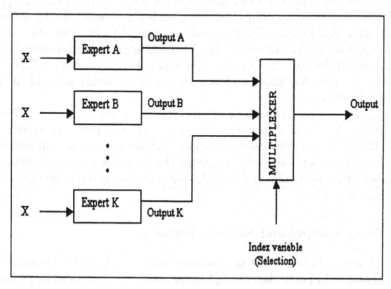

Fig. 6.7. Multiplexer used as a logic function

6.3 Fuzzy Logic Methods for Response Integration

The importance of this part of the architecture for pattern recognition is due to the high dimensionality of this type of problems. As a consequence in pattern recognition is good alternative to consider a modular approach. This has the advantage of reducing the time required of learning and it also increases accuracy. In our case, we consider dividing the images of a human face in three different regions. We also divide the fingerprint into three parts, and applying a modular structure for achieving pattern recognition.

In the literature we can find several methods for response integration, that have been researched extensively, which in many cases are based on statistical decision methods. We will mention briefly some of these methods of response integration, in particular the ones based on fuzzy logic. The idea of using these types of methods, is that the final decision takes into account all of the different kinds of information available about the human face and fingerprint. In particular, we consider aggregation operators, and the fuzzy Sugeno integral (Sugeno, 1974).

Yager mentions in his work (Yager, 1999), that fuzzy measures for the aggregation criteria of two important classes of problems. In the first type of problems, we have a set $Z = \{z_1, z_2, \ldots, z_n\}$ of objects, and it is desired to select one or more of these objects based on the satisfaction of certain criteria. In this case, for each $z_i \in Z$, it is evaluated $D(z_i) = G(A_i(z_i), \ldots, A_j(z_i))$, and then an object or objects are selected based on the value of G. The problems that fall within this structure are the multi-criteria decision problems, search in databases and retrieving of documents.

In the second type of problems, we have a set $G = \{G_1, G_2, \ldots, G_q\}$ of aggregation functions and object z. Here, each G_k corresponds to different possible identifications of object z, and our goal is to find out the correct identification of z. For achieving this, for each aggregation function G, we obtain a result for each $z, Dk(z) = Gk(A1(z), A2(z), \ldots, An(z))$. Then we associate to z the identification corresponding to the larger value of the aggregation function.

A typical example of this type of problems is pattern recognition. Where A_j corresponds to the attributes and $A_j(z)$ measures the compatibility of z with the attribute. Medical applications and fault diagnosis fall into this type of problems. In diagnostic problems, the A_j corresponds to symptoms associated with a particular fault, and G_k captures the relations between these faults.

6.3.1 Fuzzy Integral and Sugeno Measures

Fuzzy integrals can be viewed as non-linear functions defined with respect to fuzzy measures. In particular, the "gλ-fuzzy measure" introduced by (Sugeno, 1974) can be used to define fuzzy integrals. The ability of fuzzy integrals to combine the results of multiple information sources has been mentioned in previous works.

Definition 6.1. *A function of sets g:* $2^x - (0.1)$ *is called a fuzzy measure if:*

(1) $g(0) = 0 \ g(x) = 1$
(2) $g(A) \leq g(B)$ *if* $A \subset B$
(3) *if* $\{A_i\}i^\alpha = 1$ *is a sequence of increments of the measurable set then*

$$\lim_{i \to \infty} g(A_i) = g(\lim_{i \to \infty} A_i) \tag{6.3}$$

From the above it can be deduced that g is not necessarily additive, this property is replaced by the additive property of the conventional measure.

From the general definition of the fuzzy measure, Sugeno introduced what is called "gλ-fuzzy measure", which satisfies the following additive property: For every $A, B \subset X$ and $A \cap B = \theta$,

$$g(A \cup B) = g(A) + g(B) + \lambda g(A)g(B) ,$$
$$\text{for some value of } \lambda > -1 . \tag{6.4}$$

This property says that the measure of the union of two disjunct sets can be obtained directly from the individual measures. Using the concept of fuzzy measures, Sugeno developed the concept of fuzzy integrals, which are non-linear functions defined with respect to fuzzy measures like the $g\lambda$-fuzzy measure.

Definition 6.2. *Let X be a finite set and $h : X \to [0,1]$ be a fuzzy subset of X, the fuzzy integral over X of function h with respect to the fuzzy measure g is defined in the following way,*

$$h(x) \circ g(x) = \max_{E \subseteq X}[\min(\min_{x \in E} h(x), g(E))]$$
$$= \sup_{\alpha \in [0,1]} [\min(\alpha, g(h_\alpha))] \tag{6.5}$$

where h_α is the level set α of h,

$$h_\alpha = \{x | h(x) \geq \alpha\} . \tag{6.6}$$

We will explain in more detail the above definition: $h(x)$ measures the degree to which concept h is satisfied by x. The term $\min(h_x)$ measures the degree to which concept h is satisfied by all the elements in E. The value $g(E)$ is the degree to which the subset of objects E satifies the concept measure by g. As a consequence, the obtained value of comparing these two quantities in terms of operator min indicates the degree to which E satifies both criteria g and $\min(h_x)$. Finally, operator max takes the greatest of these terms. One can interpret fuzzy integrals as finding the maximum degree of similarity between the objective and expected value.

Now let us consider a simple example of the calculation of the Sugeno fuzzy measure. Consider the set $X = \{a, b, c\}$. The fuzzy membership values are given as follows:

$$g(\{a\}) = 0.3, \; g(\{b\}) = 0.4, \; g(\{c\}) = 0.1 \; .$$

Assuming the value of $\lambda = 1$, then the Sugeno measures can be calculated as follows:

$$g(\{a, b\}) = g(\{a\}) + g(\{b\}) + \lambda g(\{a\})g(\{b\}) = 0.82$$
$$g(\{a, c\}) = g(\{a\}) + g(\{c\}) + \lambda g(\{a\})g(\{c\}) = 0.43$$
$$g(\{a, b, c\}) = g(x) = 1$$

6.3.2 Fuzzy Integrals to Combine Networks

The fuzzy integral can be regarded as an aggregation operator. Let X be a set of elements (e.g. features, sensors, classifiers). Let $h : X \rightarrow [0, 1]$, $h(x)$ denotes the confidence value delivered by element x (e.g. the class membership of data determined by a specific classifier). The fuzzy integral of h over E (a subset of X) with respect to the fuzzy measure g can be calculated as follows:

$$\int h(x)^{\circ}g = \sup[\alpha \wedge g(E \cap H_{\alpha})] \tag{6.7}$$

with $H_{\alpha} = \{x | h(x) \geq \alpha\}$. In image processing we always have finite sets of elements. The computation of the fuzzy integral is easy for a finite set $X = \{x_1, x_2, \ldots, x_n\}$. If the elements are sorted so that $h(x_i)$ is a descending function, $h(x_1) \geq h(x_2) \geq \ldots \geq h(x_n)$, then the fuzzy integral can be calculated as follows:

$$e = \max_{i=j}^{n}[\min(h(x_i), g(A_i))] \tag{6.8}$$

with $A_i = \{x_1, x_2, \ldots, x_n\}$. We can note that when g is a $g\lambda$-fuzzy measure, the values of $g(A_i) = g(\{x_i\}) = g^1$. For finding the λ value we need to solve the following algebraic equation:

$$\lambda + 1 = \prod_{i=1}^{n}(1 + \lambda g^i) \tag{6.9}$$

where $\lambda \in (-1, +\infty)$, and $\lambda \neq 0$. This equation can be solved easily by finding the only root greater than -1 for the polynomial of degree $(n - 1)$.

Let $\Omega = \{w_1, w_2, \ldots, w_c\}$ be a set of classes of interest for a particular application. W_i can be even considered as a set of classes. Let $Y = (y_1, y_2, \ldots, y_n)$

be a set of neural networks, and A the object to be recognized. Let $h_k : Y \to [0, 1]$ be a partial evaluation of object A for each class w_k, which means that, $h_k(y_i)$ is an indication about the certainty of classifying object A as belonging to class wk using the neural network y_i, where 1 indicates complete certainty about object A really belonging to class wk and 0 means complete certainty about A not belonging to class w_k.

To each y_i corresponds a degree of importance g^I, that indicates how important is y_i in the recognition of class wk. This importance can be assigned in a subjective manner by a human expert, or can be inferred from the same data. The g^i's define a fuzzy importance mapping.

The λ value is calculated from (6.9) and as a consequence $g\lambda$-fuzzy measure is obtained. Finally, the class w_k with the highest fuzzy integral value is selected as the output.

$$g(A_i) = g^i + g(A_i - 1) + \lambda g^i g(A_i - 1) \tag{6.10}$$

Let us now consider a simple example of the calculation of the fuzzy Sugeno integral. Assuming that h is given as follows:

$$h(x) = 0.9 \text{ for } x = a, h(x) = 0.6 \text{ for } x = b, \text{and } h(x) = 0.3 \text{ for } x = c . \tag{6.11}$$

The fuzzy Sugeno integral with respect to the Sugeno fuzzy measure defined above can be calculated as follows:

$$E = \max[\min(0.9, 0.3), \min(0.6, 0.82), \min(0.3, 1)]$$
$$= \max[0.3, 0.6, 0.3] = 0.6$$

We can give an interpretation to this value as the one indicating the class selected for the corresponding problem.

6.4 Example of the Application of a Gating Modular Network

We show the feasibility of the proposed modular neural network approach with a problem of non-linear system identification. We consider the non-linear function given by the following mathematical equation:

$$y = 2e^{(-0.2X)}|\sin(x)| \tag{6.12}$$

This non-linear function can be considered as unknown in a problem of system identification. We will assume that we have a set of points for this function, and we will use the set of points as training data for the neural networks.

The graph of this non-linear function is shown in Fig. 6.8. From this figure, we can appreciate that this function has three different regions. For this reason, we can use this function to test the modular approach, i.e. one neural network module for each region can be used.

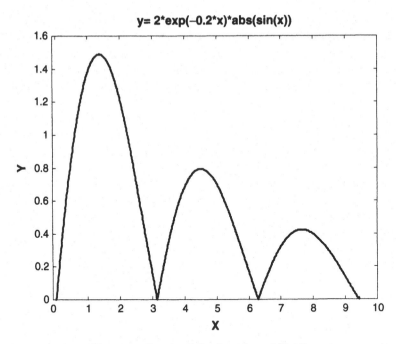

Fig. 6.8. Graph of the function of (6.12)

The idea of using one neural network (module) for each of the three different regions of the non-linear function is illustrated in Fig. 6.9.

We used as range of values for the function the closed interval [0, 9.45] with a 0.01 step size, which gives us 945 points to use as training data for the three modules. The data is divided in the three modules as follows:

Module 1: points from 0 to 3.15
Module 2: points from 3.15 to 6.30
Module 3: points from 6.30 to 9.45

The idea behind this data partition is that learning will be easier in each of the modules, i.e. a simple NN can learn more easily the behavior of the function in one of the regions. We used three-layer feed-forward neural networks for each of the modules with the Levenberg-Marquardt training algorithm. We show in Fig. 6.10 the general architecture of the MNN used in this example.

We show in Fig. 6.11 the topology of the final modules of the neural network (obtained after experimenting with the number of nodes of the hidden layer) for the problem of function identification. As we can appreciate, from this figure, module 2 is the smallest one and module 3 is the largest one. The resulting architecture of the MNN is a particular architecture with different sizes of the modules (neural networks). In general, it is difficult to know a-priori the specific architecture of the modules and many times trial-and-error is used, but it is also possible to use evolutionary techniques (like genetic

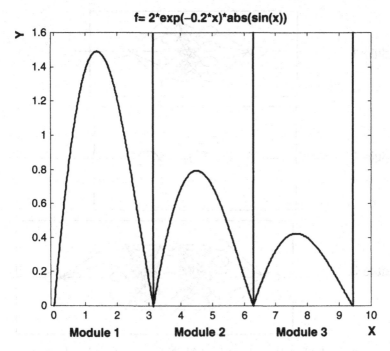

Fig. 6.9. Modular approach to function identification

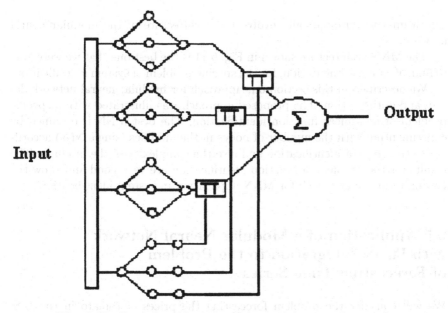

Fig. 6.10. General architecture of the modular neural network

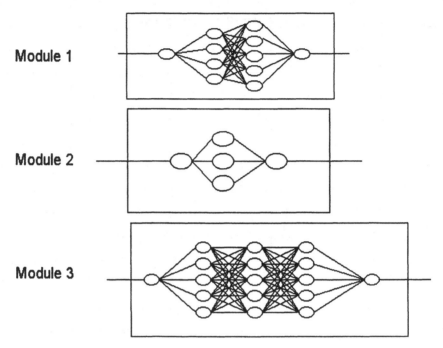

Fig. 6.11. Topology of the final MNN for function identification

algorithms) to automatically evolve the architecture of the modular neural network.

The MNN architecture shown in Fig. 6.11 is the best one that we were able to find, after experimentation, for the specific problem of system identification.

We described in this section our approach for modular neural network design and optimization. The proposed approach was illustrated with a specific problem of non-linear function identification. The best MNN is obtained by experimenting with the number of nodes in the modules (single NNs) according to the error of identification and also the complexity of the modules. The results for the problem of function identification are very good and show the feasibility of the approach for MNN application in system identification.

6.5 Application of a Modular Neural Network with Fuzzy Integration to the Problem of Forecasting Time Series

We will consider the problem forecasting the prices of tomato in the U.S. market. The time series for the prices of this consumer good shows very complicated dynamic behavior, and for this reason it is interesting to analyze and predict the future prices for this good. We show in Fig. 6.12 the time series

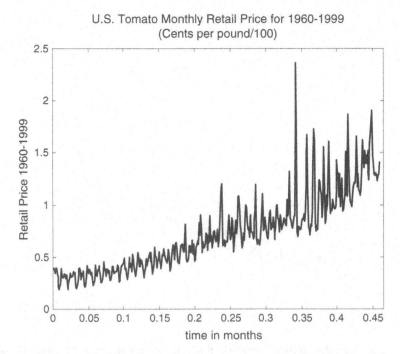

Fig. 6.12. Prices in US Dollars of tomato from January 1960 to December 1999

of monthly tomato prices in the period of 1960 to 1999, to give an idea of the complex dynamic behavior of this time series.

We will apply both the modular and monolithic neural network approach and also the linear regression method to the problem of forecasting the time series of tomato prices. Then, we will compare the results of these approaches to select the best one for forecasting.

We describe, in this section, the experimental results obtained by using neural networks to the problem of forecasting tomato prices in the U.S. Market. We show results of the application of several architectures and different learning algorithms to decide on the best one for this problem. We also compare at the end the results of the neural network approach with the results of linear regression models, to measure the difference in forecasting power of both methodologies.

First, we will describe the results of applying modular neural networks to the time series of tomato prices. We used the monthly data from 1960 to 1999 for training a Modular Neural Network with four Modules, each of the modules with 80 neurons and one hidden layer. We show in Fig. 6.13 the result of training the modular neural network with this data. In Fig. 6.13, we can appreciate how the modular neural network approximates very well the real time series of tomato prices over the relevant period of time.

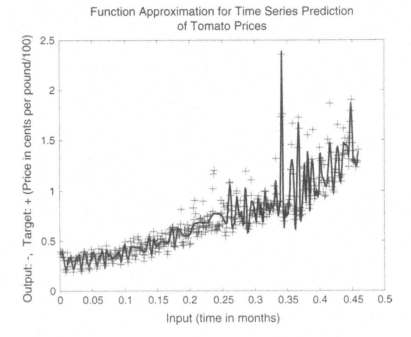

Fig. 6.13. Modular Neural network for tomato prices with the Levenberg-Marquardt algorithm

We have to mention that the results shown in Fig. 6.13 are for the best modular neural network that we were able to find for this problem by experimentation. We show in Fig. 6.14 the comparison between several of the modular neural networks that we tried in our experiments. From Fig. 6.14 we can appreciate that the modular neural network with one time delay and Leverberg-Marquardt training algorithm is the one that fits best the data and for this reason is the one selected.

We show in Fig. 6.15 the comparison of the best monolithic network against the best modular neural network. The modular network clearly fits better the real data of the problem. The explanation for these results is that clearly the MNN is able to combine in the best way possible the outputs of 4 neural networks.

We summarize the above results using neural networks, and the results of using linear regression models in Table 6.1. From Table 6.1 we can see very clearly the advantage of using modular neural networks for simulating and forecasting the time series of prices. Also, we can conclude from this table that the Levenberg-Marquardt (LM) training algorithm is better than backpropagation with momentum. Of course, the reason for this may be that the Levenberg-Marquardt training algorithm, having a variable learning rate is able to adjust to the complicated behavior of the time series.

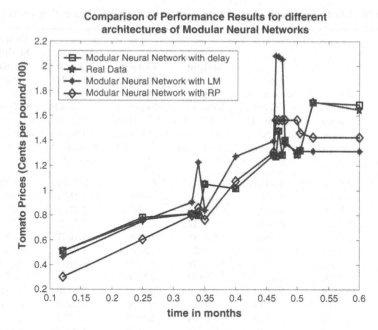

Fig. 6.14. Comparison of performance results for several modular neural networks

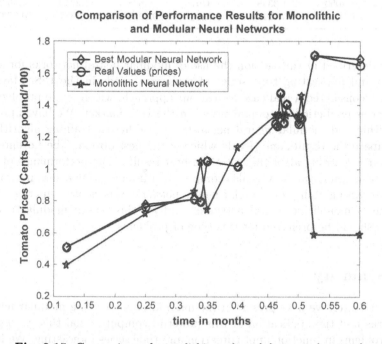

Fig. 6.15. Comparison of monolithic and modular neural networks

Table 6.1. Comparison of performance results for several neural networks

Input T Values (time in months)	Output Real Values (price cents per pound/100)	Output Monolithic Neural Network 80 50 Nodes 2 Layers SSE = 16.98	Output Modular Neural Network 4 Modules 100 Nodes 1 layer LM SSE = 3.95	Output Modular Neural Network 4 Modules 80 50 Nodes 2 layer RP SSE = 4.65	Output Modular Neural Network 4 Modules 80 Nodes 1 layer LM SSE = 5.45	Output Modular Neural Network 4 Modules 80 Nodes 1 layer LM with delay SSE = 0.5×10^{-9}
0.1210	0.5100	0.3953	0.4653	0.3021	0.4711	0.5100
0.5040	1.3180	1.4559	1.3137	1.4612	1.1659	1.3201
0.4610	1.2840	1.3349	1.3987	1.3083	1.3525	1.2840
0.5250	1.7110	0.5885	1.3137	1.4280	1.1629	1.7100
0.6000	1.6500	0.5885	1.3137	1.4280	1.1627	1.6896
0.4000	1.0160	1.1335	1.2724	1.07112	1.2852	1.0153
0.5000	1.2860	1.3349	1.3137	1.5684	1.1682	1.2928
0.3500	1.0510	0.7440	0.8420	0.7659	0.7631	1.0510
0.3300	0.8120	0.8569	0.9019	0.8002	0.8875	0.8120
0.3400	0.7970	1.0448	1.2251	0.8596	0.9412	0.7970
0.4650	1.2720	1.3349	2.0814	1.5639	1.3655	1.2720
0.4700	1.4760	1.3349	2.0795	1.5678	1.3208	1.4758
0.4750	1.2870	1.3349	2.0508	1.5678	1.2667	1.2870
0.4800	1.4050	1.3349	1.3707	1.5678	1.2252	1.4045
0.2500	0.7640	0.7219	0.7518	0.6056	0.7387	0.7785

We described in this section the use of modular neural networks for simulation and forecasting time series of consumer goods in the U.S. Market. We have considered a real case to test our approach, which is the problem of time series prediction of tomato prices in the U.S. market. We have applied monolithic and modular neural networks with different training algorithms to compare the results and decide which is the best option. The Levenberg-Marquardt learning algorithm gave the best results. The performance of the modular neural networks was also compared with monolithic neural networks. The forecasting ability of modular neural networks was clearly superior, than the one of monolithic neural networks, and for this reason modular neural models should be preferred for this type of problems.

6.6 Summary

In this chapter, we have presented the main ideas underlying modular neural networks and the application of this powerful computational theory to general problems in function approximation and time series forecasting. We have discussed in some detail the different modular neural network and ensemble architectures, different methods of response integration, and the application of fuzzy logic techniques for achieving response integration in modular neural networks. In the following chapters, we will show how modular neural network models (in conjunction with other techniques) can be applied to solve real world complex problems in intelligent pattern recognition. This chapter will serve as a basis for the new hybrid intelligent methods that will be described in the chapters at the end of this book.

7

Evolutionary Computing for Architecture Optimization

This chapter introduces the basic concepts and notation of evolutionary algorithms, which are basic search methodologies that can be used for modelling and simulation of complex non-linear dynamical systems. Since these techniques can be considered as general purpose optimization methodologies, we can use them to find the mathematical model which minimizes the fitting errors for a specific problem. On the other hand, we can also use any of these techniques for simulation if we exploit their efficient search capabilities to find the appropriate parameter values for a specific mathematical model. We also describe in this chapter the application of genetic algorithms to the problem of finding the best neural network or fuzzy system for a particular problem. We can use a genetic algorithm to optimize the weights or the architecture of a neural network for a particular application. Alternatively, we can use a genetic algorithm to optimize the number of rules or the membership functions of a fuzzy system for a specific problem. These are two important application of genetic algorithms, which will be used in later chapters to design intelligent systems for pattern recognition in real world applications.

Genetic algorithms have been used extensively for both continuous and discrete optimization problems (Jang, Sun & Mizutani, 1997). Common characteristics of these methods are described next.

- *Derivative freeness*: These methods do not need functional derivative information to search for a set of parameters that minimize (or maximize) a given objective function. Instead they rely exclusively on repeated evaluations of the objective function, and the subsequent search direction after each evaluation follows certain heuristic guidelines.
- *Heuristic guidelines*: The guidelines followed by these search procedures are usually based on simple intuitive concepts. Some of these concepts are motivated by so-called nature's wisdom, such as the evolution.
- *Flexibility*: Derivative freeness also relieves the requirement for differentiable objective functions, so we can use as complex an objective function as a specific application might need, without sacrificing too much in extra coding

Patricia Melin and Oscar Castillo: *Hybrid Intelligent Systems for Pattern Recognition Using Soft Computing*, StudFuzz **172**, 131–168 (2005)
www.springerlink.com

and computation time. In some cases, an objective function can even include the structure of a data-fitting model itself, which may be a fuzzy model.

- *Randomness*: These methods are stochastic, which means that they use random number generators in determining subsequent search directions. This element of randomness usually gives rise to the optimistic view that these methods are "global optimizers" that will find a global optimum given enough computing time. In theory, their random nature does make the probability of finding an optimal solution nonzero over a fixed amount of computation time. In practice, however, it might take a considerable amount of computation time.
- *Analytic opacity*: It is difficult to do analytic studies of these methods, in part because of their randomness and problem-specific nature. Therefore, most of our knowledge about them is based on empirical studies.
- *Iterative nature*: These techniques are iterative in nature and we need certain stopping criteria to determine when to terminate the optimization process. Let K denote an iteration count and f_k denote the best objective function obtained at count k; common stopping criteria for a maximization problem include the following:
  (1) Computation time: a designated amount of computation time, or number of function evaluations and/or iteration counts is reached.
  (2) Optimization goal: f_k is less than a certain preset goal value.
  (3) Minimal improvement: $f_k - f_{k-1}$ is less than a preset value.
  (4) Minimal relative improvement: $(f_k - f_{k-1})/f_{k-1}$ is less than a preset value.

Evolutionary algorithms (EAs), in general, and genetic algorithms (GAs), in particular, have been receiving increasing amounts of attention due to their versatile optimization capabilities for both continuous and discrete optimization problems. Moreover, both of them are motivated by so-called "nature's wisdom": EAs are based on the concepts of natural selection and evolution; while GAs consider genetic information in a more simple binary form.

7.1 Genetic Algorithms

Genetic algorithms (GAs) are derivative-free optimization methods based on the concepts of natural selection and evolutionary processes (Goldberg, 1989). They were first proposed and investigated by John Holland at the University of Michigan (Holland, 1975). As a general-purpose optimization tool, GAs are moving out of academia and finding significant applications in many areas. Their popularity can be attributed to their freedom from dependence on functional derivatives and their incorporation of the following characteristics:

- GAs are parallel-search procedures that can be implemented on parallel processing machines for massively speeding up their operations.

- GAs are applicable to both continuous and discrete (combinatorial) optimization problems.
- GAs are stochastic and less likely to get trapped in local minima, which inevitably are present in any optimization application.
- GAs' flexibility facilitates both structure and parameter identification in complex models such as fuzzy inference systems or neural networks.

GAs encode each point in a parameter (or solution) space into a binary bit string called a "chromosome", and each point is associated with a "fitness value" that, for maximization, is usually equal to the objective function evaluated at the point. Instead of a single point, GAs usually keep a set of points as a "population", which is then evolved repeatedly toward a better overall fitness value. In each generation, the GA constructs a new population using "genetic operators" such as crossover and mutation; members with higher fitness values are more likely to survive and to participate in mating (crossover) operations. After a number of generations, the population contains members with better fitness values; this is analogous to Darwinian models of evolution by random mutation and natural selection. GAs and their variants are sometimes referred to as methods of "population-based optimization" that improve performance by upgrading entire populations rather than individual members. Major components of GAs include encoding schemes, fitness evaluations, parent selection, crossover operators, and mutation operators; these are explained next.

Encoding schemes: These transform points in parameter space into bit string representations. For instance, a point (11, 4, 8) in a three-dimensional parameter space can be represented as a concatenated binary string:

$$\underbrace{1011}_{11} \ \underbrace{0100}_{4} \ \underbrace{1000}_{8}$$

in which each coordinate value is encoded as a "gene" composed of four binary bits using binary coding. other encoding schemes, such as gray coding, can also be used and, when necessary, arrangements can be made for encoding negative, floating-point, or discrete-valued numbers. Encoding schemes provide a way of translating problem-specific knowledge directly into the GA framework, and thus play a key role in determining GAs' performance. Moreover, genetic operators, such as crossover and mutation, can and should be designed along with the encoding scheme used for a specific application.

Fitness evaluation: The first step after creating a generation is to calculate the fitness value of each member in the population. For a maximization problem, the fitness value f_i of the ith member is usually the objective function evaluated at this member (or point). We usually need fitness values that are positive, so some kind of monotonical scaling and/or translation may by necessary if the objective function is not strictly positive. Another approach is to use the rankings of members in a population as their fitness values. The advantage of this is that the objective function does not need to be accurate, as long as it can provide the correct ranking information.

Selection: After evaluation, we have to create a new population from the current generation. The selection operation determines which parents participate in producing offspring for the next generation, and it is analogous to "survival of the fittest" in natural selection. Usually members are selected for mating with a selection probability proportional to their fitness values. The most common way to implement this is to set the selection probability equal to:

$$f_i \left/ \sum_{k=1}^{k=n} f_k \right. ,$$

where n is the population size. The effect of this selection method is to allow members with above-average fitness values to reproduce and replace members with below-average fitness values.

Crossover: To exploit the potential of the current population, we use "crossover" operators to generate new chromosomes that we hope will retain good features from the previous generation. Crossover is usually applied to selected pairs of parents with a probability equal to a given "crossover rate". "One-point crossover" is the most basic crossover operator, where a crossover point on the genetic code is selected at random and two parent chromosomes are interchanged at this point. In "two-point crossover", two crossover points are selected and the part of the chromosome string between these two points is then swapped to generate two children. We can define n-point crossover similarly. In general, $(n-1)$-point crossover is a special case of n-point crossover. Examples of one-and two-point crossover are shown in Fig. 7.1.

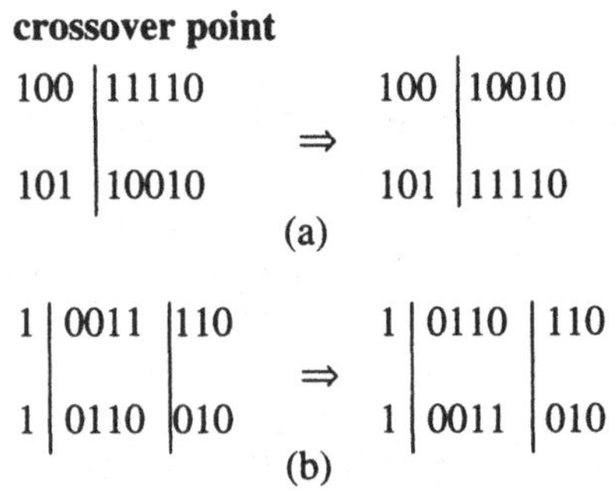

crossover point

| 100 | 11110 | | 100 | 10010 |
| 101 | 10010 | $\Rightarrow$ | 101 | 11110 |

(a)

| 1 | 0011 | 110 | | 1 | 0110 | 110 |
| 1 | 0110 | 010 | $\Rightarrow$ | 1 | 0011 | 010 |

(b)

Fig. 7.1. Crossover operators: (a) one-point crossover; (b) two-point crossover

Mutation: Crossover exploits current gene potentials, but if the population does not contain all the encoded information needed to solve a particular problem, no amount of gene mixing can produce a satisfactory solution. For this reason, a "mutation" operator capable of spontaneously generating new chromosomes is included. The most common way of implementing mutation is to flip a bit with a probability equal to a very low given "mutation rate". A mutation operator can prevent any single bit from converging to a value throughout the entire population and, more important, it can prevent the population from converging and stagnating at any local optima. The mutation rate is usually kept low so good chromosomes obtained from crossover are not lost. If the mutation rate is high (above 0.1), GA performance will approach that of a primitive random search. Figure 7.2 provides an example of mutation.

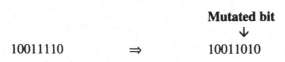

Fig. 7.2. Mutation operator

In the natural evolutionary process, selection, crossover, and mutation all occur in the single act of generating offspring. Here we distinguish them clearly to facilitate implementation of and experimentation with GAs.

Based on the aforementioned concepts, a simple genetic algorithm for maximization problems is described next.

Step 1: Initialize a population with randomly generated individuals and evaluate the fitness value of each individual.

Step 2: Perform the following operations:
  (a) Select two members from the population with probabilities proportional to their fitness values.
  (b) Apply crossover with a probability equal to the crossover rate.
  (c) Apply mutation with a probability equal to the mutation rate.
  (d) Repeat (a) to (d) until enough members are generated to form the next generation.

Step 3: Repeat steps 2 and 3 until a stopping criterion is met.

Figure 7.3 is a schematic diagram illustrating how to produce the next generation from the current one.

Lets consider a simple example to illustrate the application of the basic genetic algorithm. Will consider the maximization of the "peaks" function, which is given by the following equation:

$$Z = f(x,y) = 3(1-x)^2 e^{-x2-(y+1)2} - 10(x/5 - x^3 - y^5)e^{-x2-y2}$$
$$- (1/3)e^{-(x+1)2-y2} . \tag{7.1}$$

The surface plot of this function is shown in Fig. 7.4. To use Gas to find the maximum of this function, we first confine the search domain to the square $[-3,3] \times [-3,3]$. We use 8-bit binary coding for each variable. Each generation in our GA implementation contains 30 points or individuals. We use a

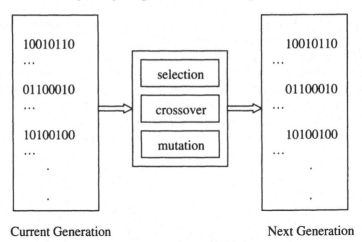

Fig. 7.3. Producing the next generation in GAs

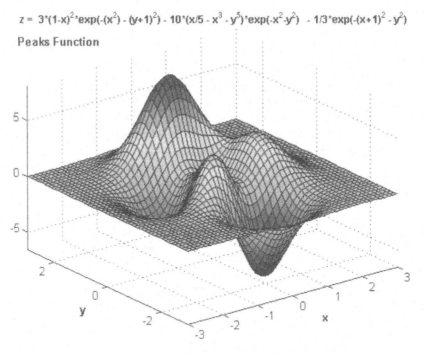

Fig. 7.4. Surface plot of the "peaks" function

simple one-point crossover scheme with crossover rate equal to 0.9. We choose uniform mutation with mutation rate equal to 0.05. Figure 7.5 shows a plot of the best, average, and poorest values of the objective function across 30 generations. Figure 7.6 shows the contour plot of the "peaks" function with the final population distribution after 30 generations. We can appreciate from these figures how the GA is able to find the maximum value of 8.

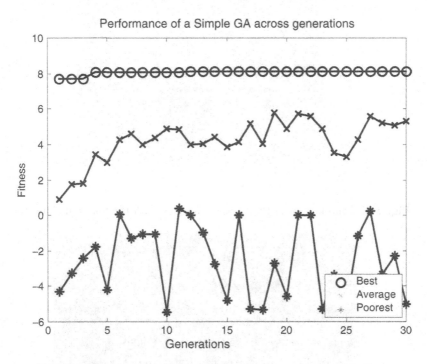

Fig. 7.5. Performance of the genetic algorithm across generations

Now lets consider a more complicated example. We will consider the Rosenbrock's valley, which is a classic optimization problem. The global optimum is inside a long, narrow, parabolic shaped flat valley. To find the valley is trivial, however convergence to the global optimum is difficult and hence this problem has been repeatedly used in assess the performance of optimization algorithms. In this case, the function is given by the following equation

$$f_2 = \sum_i^{n-1} 100 \left(x_2 - x_1^2\right)^2 + (1 - x_1)^2 \, -2 \le x_i \le 2$$

$$\text{global minimum: } x_i = 1 \quad f(x) = 0 \,. \tag{7.2}$$

We show in Fig. 7.7 the plot of Rosenbrock's valley function. We will apply a simple genetic algorithm with the same parameters as in the previous example, i.e. same population size, mutation rate, and crossover rate. The only

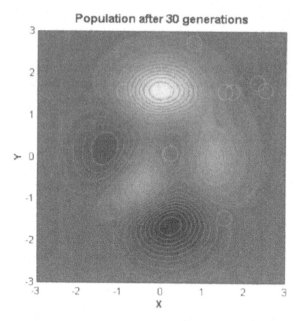

Fig. 7.6. Contour plot of the "peaks" function with the final population

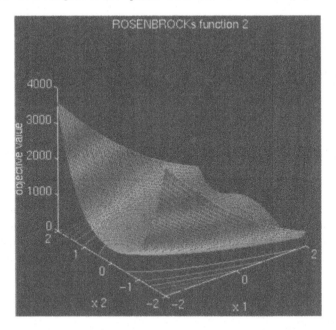

Fig. 7.7. Plot of the Rosenbrock's valley

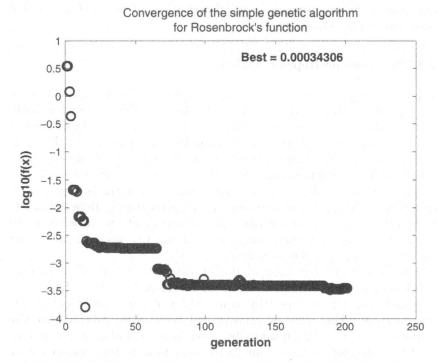

Fig. 7.8. Performance of the genetic algorithm for Rosenbrock's valley

parameter we will change is the maximum number of generations, which will now be of 250. Figure 7.8 shows the performance of the simple genetic algorithm for Rosenbrock's function. We can notice from this figure that the GA is able to converge in about 200 generations to a best value of 0.00034306. We have to mention that the computer program to obtain these results was implemented in the MATLAB programming language.

7.2 Modifications to Genetic Algorithms

The GA mechanism is neither governed by the use of differential equations nor does it behave like a continuous function. However, it possesses the unique ability to search and optimize a solution for complex system, where other mathematical oriented techniques may have failed to compile the necessary design specifications. Due to its evolutionary characteristics, a standard GA may not be flexible enough for a practical application, and an engineering insight is always required whenever a GA is applied. This becomes more apparent where the problem to be tackled is complicated, multi-tasking and conflicting. Therefore, a means of modifying the GA structure is sought in order to meet the design requirements. There are many facets of operational modes that can be introduced.

7.2.1 Chromosome Representation

The problem to be tackled varies from one to the other. The coding of chromosome representation may vary according to the nature of the problem itself. In general, the bit string encoding (Holland, 1975) is the most classic method used by GA researchers because of its simplicity and traceability. The conventional GA operations and theory (scheme theory) are also developed on the basis of this fundamental structure. Hence, this representation is adopted in many applications. However, one minor modification can be suggested in that a Gray code may be used instead of the binary coding. Hollstien (1971) investigated the use of GA for optimizing functions of two variables based on a Gray code representation, and discovered that this works slightly better than the normal binary representation.

More recently, a direct manipulation of real-value chromosomes (Janikow & Michalewicz, 1991; Wright, 1991) raised considerable interest. This representation was introduced especially to deal with real parameter problems. The work currently taking place by Janikow and Michalewicz indicates that the floating-point representation would be faster in computation and more consistent from the basis of run-to-run. At the same time, its performance can be enhanced by special operators to achieve high accuracy (Michalewicz, 1996).

7.2.2 Objective Functions and Fitness

An objective function is a measuring mechanism that is used to evaluate the status of a chromosome. This is a very important link to relate the GA and the system concerned. Since each chromosome is individually going through the same calculations, the range of this value varies from one chromosome to another. To maintain uniformity, the objective value O is mapped into a fitness value with a map Ψ where the domain of F is usually greater than zero.

$$\Psi : O \to F \tag{7.3}$$

Linear Scaling

The fitness value f_i of chromosome i has a linear relationship with the objective value o_i as

$$f_i = ao_i + b \tag{7.4}$$

where a and b are chosen to enforce the equality of the average objective value and the average fitness value, and cause maximum scaled fitness to be a specified multiple of the average fitness.

This method can reduce the effect of genetic drift by producing an extraordinarily good chromosome. However, it may introduce a negative fitness value, which must be avoided. Hence, the choice of a and b are dependent on the knowledge of the range of the objective values.

Sigma Truncation

This method avoids the negative fitness value and incorporates the problem dependent information into the scaling mechanism. The fitness value f_i of chromosome i is calculated according to

$$f_i = o_i - (\tilde{o} - c\sigma) \tag{7.5}$$

where c is small integer, $\tilde{o}$ is the mean of the objective values, σ is the standard deviation in the population.

To prevent negative values of f, any negative result $f < 0$ is arbitrarily set to zero. Chromosomes whose fitness values are less than c (a small integer from the range 1 and 5) standard deviations from the average fitness value are not selected.

Power Law Scaling

The actual fitness value is taken as a specific power of the objective value, o_i

$$Fi = o_i^k \tag{7.6}$$

where k is in general problem dependent or even varying during the run (Gillies, 1985).

Ranking

There are other methods that can be used such as the Ranking scheme (Baker, 1987). The fitness values do not directly relate to their corresponding objective values, but to the ranks of the objective values.

Using this approach can help the avoidance of premature convergence and speed up the search when the population approaches convergence. On the other hand, it requires additional overheads in the in the GA computation for sorting chromosomes according to their objective values.

7.2.3 Selection Methods

To generate good offspring, a good parent selection mechanism is necessary. This is a process used for determining the number of trials for one particular individual used in reproduction. The chance of selecting one chromosome as a parent should be directly proportional to the number of offspring produced.

Baker (1987) presented three measures of performance of the selection methods: Bias, Spread, and Efficiency. *Bias* defines the absolute difference between individuals in actual and expected probability for selection. Optimal zero bias is achieved when an individual's probability equals its expected number of trials. *Spread* is a range in the possible number of trials that an individual may achieve. If $g(i)$ is the actual number of trials due to each individual i, then the "minimum spread" is the smallest spread that theoretically permits zero bias, i.e.

$$G(i) \in \{\lfloor et(i) \rfloor, \lceil et(i) \rceil\} \tag{7.7}$$

where $et(i)$ is the expected number of trials of individual i, $\lfloor et(i) \rfloor$ is the floor and $\lceil et(i) \rceil$ is the ceiling. Thus the spread of a selection method measures its consistency. *Efficiency* is related to the overall time complexity of the algorithms.

The selection method should thus be achieving a zero bias whilst maintaining a minimum spread and not contributing to an increased time complexity of the GA.

Many selection techniques employ the "roulette wheel mechanism". The basic roulette wheel selection method is a stochastic sampling with replacement (SSR) technique. The segment size and selection probability remain the same throughout the selection phase and the individuals are selected according to the above procedures. SSR tends to give zero bias but potentially inclines to a spread that is unlimited.

Stochastic Sampling with Partial Replacement (SSPR) extends upon SSR by resizing a chromosome's segment if it is selected. Each time a chromosome is selected, the size of its segment is reduced by a certain factor. If the segment size becomes negative, then it is set to zero. This provides an upper bound on the spread of $\lfloor et(i) \rfloor$ but with a zero lower bound and a higher bias. The roulette wheel selection methods can generally be implemented with a time complexity of the order of $NlogN$ where N is the population size.

Stochastic Universal Sampling (SUS) is another single-phase sampling method with minimum spread, zero bias and the time complexity is in the order of N (Baker, 1987). SUS uses an N equally spaced pointer, where N is the number of selections required. The population is shuffled randomly and a single random number in the range $[0, F_{sum}/N]$ is generated, ptr, where F_{sum} is the sum of the individuals' fitness values. An individual is thus guaranteed to be selected a minimum of $\lfloor et(i) \rfloor$ times and no more than $\lceil et(i) \rceil$, thus achieving minimum spread. In addition, as individuals are selected entirely based on their position in the population, SUS has zero bias.

7.2.4 Genetic Operations

Crossover

Although the one-point crossover method was inspired by biological processes, it has one major drawback in that certain combinations of schema cannot be combined in some situations (Michalewicz, 1996).

A multi-point crossover can be introduced to overcome this problem. As a result, the performance of generating offspring is greatly improved. Another approach is the uniform crossover. This generates offspring from the parents, based on a randomly generated crossover mask. The resulting offspring contain a mixture of genes from each parent. The number of effective crossing points is not fixed, but will be averaged at $L/2$ (where L is the chromosome length).

The preference for using which crossover techniques is still arguable. However, De Jong (1975) concluded that a two-point crossover seemed to be an optimal number for multi-point crossover. Since then, this has been contradicted by Spears and De Jong (1991) as a two-point crossover could perform poorly if the population has largely being converged because of any reduced crossover productivity. A general comment was that each of these crossover operators was particularly useful for some classes of problems and quite poor for others, and that the one-point crossover was considered a "loser" experimentally.

Crossover operations can be directly adopted into the chromosome with real number representation. The only difference would be if the string is composed of a series of real numbers instead of binary numbers.

Mutation

Originally, mutation was designed only for the binary-represented chromosome. To adopt the concept of introducing variations into the chromosome, a random mutation (Michalewicz, 1996) has been designed for a real number chromosome:

$$g = g + \psi(\mu, \sigma) \tag{7.8}$$

where g is the real number gene; ψ is a random function which may be Gaussian or normally distributed; μ is the mean and σ is the variance of the random function.

Operational Rates Settings

The choice of an optimal probability operation rate for crossover and mutation is another controversial debate for both analytical and empirical investigations. The increase of crossover probability would cause the recombination of building blocks to rise, and at the same time, it also increases the disruption of good chromosomes. On the other hand, should the mutation probability increase, this would transform the genetic search into a random search, but would help to reintroduce the lost genetic material.

7.2.5 Parallel Genetic Algorithm

Considering that the GA already possesses an intrinsic, parallelism architecture, there is no extra effort to construct a parallel computational framework. Rather, the GA can be fully exploited in its parallel structure to gain the required speed for practical applications.

There are a number of GA-based parallel methods to enhance the computational speed (Cantú-Paz, 1995). The methods of parallelization can be classified as Global, Migration and Diffusion. These categories reflect different ways in which parallelism can be exploited in the GA as well as the nature of the population structure and recombination mechanisms used.

Global GA

Global GA treats the entire population as a single breeding mechanism. This can be implemented on a shared memory multiprocessor or distributed memory computer. On a shared memory multiprocessor, chromosomes are stored in the shared memory. Each processor accesses the particular assigned chromosome and returns the fitness values without any conflicts. It should be noted that there is some synchronization needed between generation to generation. It is necessary to balance the computational load among the processors using a dynamic scheduling algorithm.

On a distributed memory computer, the population can be stored in one processor to simplify the genetic operators. This is based on the farmer-worker architecture. The farmer processor is responsible for sending chromosomes to the worker processors for the purpose of fitness evaluation. It also collects the results from them, and applies the genetic operators for producing the next generation.

Migration GA

This is another parallel processing method for computing the GA. The migration GA divides the population into a number of sub-populations, each of which is treated as a separate breeding unit under the control of a conventional GA. To encourage the proliferation of good genetic material throughout the whole population, migration between the sub-populations occurs from time to time. The required parameters for successful migration are the "migration rate" and the "migration interval". The migration rate governs the number of individuals to be migrated. The migration interval affects the frequency of migrations. The values of these parameters are intuitively chosen rather than based on some rigorous scientific analysis. In general, the occurrence of migration is usually set at a predetermined constant interval that is governed by migration intervals. We illustrate the concept of migration in Fig. 7.9.

The migration GA is well suited to parallel implementation on Multiple Instruction Multiple Data (MIMD) machines. The architecture of hypercubes

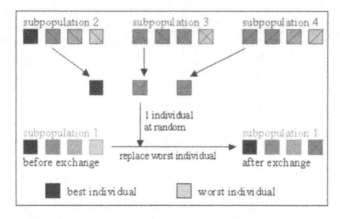

Fig. 7.9. Detailed description of the migration concept

and rings are commonly used for this purpose. Given the range of possible population topologies and migration paths between them, efficient communication networks should thus be possible on most parallel architectures. This applies to small multiprocessor platforms or even the clustering of networked workstations.

Diffusion GA

The Diffusion GA considers the population as a single continuous structure. Each individual is assigned to a geographic location on the population surface and usually placed in a two-dimensional grid. This is because of the topology of the processing element in many massively parallel computers that are constructed in this form. The individuals are allowed to breed with individuals contained in a small local neighborhood. This neighborhood is usually chosen from immediately adjacent individuals on the population surface and is motivated by the practical communication restrictions of parallel computers.

We will illustrate the ideas of parallel genetic algorithms with the well known harvest optimization problem. We will apply a migration GA to solve this optimization problem. The harvest system is a one-dimensional equation of growth with one constraint:

$$x(k+1) = ax(k) - u(k) \ k = 1, \ldots, N, \quad \text{such that } x(0) = x(N) . \quad (7.9)$$

The objective function for minimization is therefore defined as:

$$F(u) = -\sum_{k=1}^{N} [u(k)]^{1/2} . \quad (7.10)$$

We will apply a multi-population genetic algorithm with migration and real-valued representation of the individuals. We used 20 decision variables, a

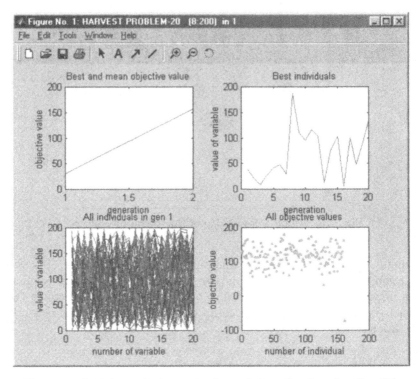

Fig. 7.10. Initial population and subpopulations for the migration GA

crossover rate of 1.0 and a mutation rate of 1/number of variables. We used 8 subpopulations with 20 individuals each, and a migration rate of 0.2. The maximum number of generations was specified at 200. We show in Fig. 7.10 the initial population of individuals. In Fig. 7.11 we show the simulation results after 200 generations of the GA.

7.3 Applications of Genetic Algorithms

The genetic algorithms described in the previous sections are very simple, but variations of these algorithms have been used in a large number of scientific and engineering problems and models (Mitchell, 1996). Some examples follow.

- Optimization: genetic algorithms have been used in a wide variety of optimization tasks, including numerical optimization and such combinatorial optimization problems as circuit layout and job-shop scheduling.
- Automatic Programming: genetic algorithms have been used to evolve computer programs for specific tasks, and to design other computational structures such as cellular automata and sorting networks.

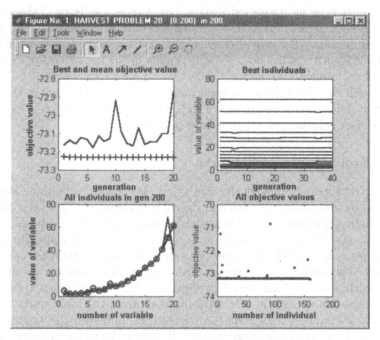

Fig. 7.11. Final population for the migration GA after 200 generations

- Machine Learning: genetic algorithms have been used for many machine learning applications, including classification and prediction tasks, such as the prediction of weather or protein structure. Genetic algorithms have also been used to evolve aspects of particular machine learning systems, such as weights for neural networks, rules for learning classifier systems or symbolic production systems, and sensors for robots.
- Economics: genetic algorithms have been used to model processes of innovation, the development of bidding strategies, and the emergence of economic markets.
- Immune Systems: genetic algorithms have been used to model various aspects of natural immune systems, including somatic mutation during an individual's lifetime and the discovery of multi-gene families during evolutionary time.
- Ecology: genetic algorithms have used to model ecological phenomena such as biological arms races, host-parasite co-evolution, symbiosis, and resource flow.
- Social Systems: genetic algorithms have been used to study evolutionary aspects of social systems, such as the evolution of social behavior in insect colonies, and, more generally, the evolution of cooperation and communication in multi-agent systems.

This list is by no means exhaustive, but it gives the flavor of the kinds of things genetic algorithms have been used for, both in problem solving and in scientific contexts. Because of their success in these and other areas, interest in genetic algorithms has been growing rapidly in the last several years among researchers in many disciplines.

We will describe bellow the application of genetic algorithms to the problem of evolving neural networks, which is a very important problem in designing the particular neural network for a problem.

7.3.1 Evolving Neural Networks

Neural Networks are biologically motivated approaches to machine learning, inspired by ideas from neuroscience. Recently, some efforts have been made to use genetic algorithms to evolve aspects of neural networks (Mitchell, 1996).

In its simplest feedforward form, a neural network is a collection of connected neurons in which the connections are weighted, usually with real-valued weights. The network is presented with an activation pattern on its input units, such as a set of numbers representing features of an image to be classified. Activation spreads in a forward direction from the input units through one or more layers of middle units to the output units over the weighted connections. This process is meant to roughly mimic the way activation spreads through networks of neurons in the brain. In a feedforward network, activation spreads only in a forward direction, from the input layer through the hidden layers to the output layer. Many people have also experimented with "recurrent" networks, in which there are feedback connections between layers.

In most applications, the neural network learns a correct mapping between input and output patterns via a learning algorithm. Typically the weights are initially set to small random values. Then a set of training inputs is presented sequentially to the network. In the backpropagation learning procedure, after each input has propagated through the network and an output has been produced, a "teacher" compares the activation value at each output unit with the correct values, and the weights in the network are adjusted in order to reduce the difference between the network's output and the correct output. This type of procedure is known as "supervised learning", since a teacher supervises the learning by providing correct output values to guide the learning process.

There are many ways to apply genetic algorithms to neural networks. Some aspects that can be evolved are the weights in a fixed network, the network architecture (i.e., the number of neurons and their interconnections can change), and the learning rule used by the network.

Evolving Weights in a Fixed Network

David Montana and Lawrence Davis (1989) took the first approach of evolving the weights in a fixed network. That is, Montana and Davis were using the genetic algorithm instead of backpropagation as a way of finding a good set of weights for a fixed set of connections. Several problems associated with the backpropagation algorithm (e.g., the tendency to get stuck at local minima, or the unavailability of a "teacher" to supervise learning in some tasks) often make it desirable to find alternative weight training schemes.

Montana and Davis were interested in using neural networks to classify underwater sonic "lofargrams" (similar to spectrograms) into two classes: "interesting" and "not interesting". The networks were to be trained from a database containing lofargrams and classifications made by experts as to whether or not a given lofargram is "interesting". Each network had four input units, representing four parameters used by an expert system that performed the same classification. Each network had one output unit and two layers of hidden units (the first with seven units and the second with ten units). The networks were fully connected feedforward networks. In total there were 108 weighted connections between units. In addition, there were 18 weighted connections between the non-input units and a "threshold unit" whose outgoing links implemented the thresholding for each of the non-input units, for a total of 126 weights to evolve.

The genetic algorithm was used as follows. Each chromosome was a list of 126 weights. Figure 7.12 shows (for a much smaller network) how the encoding was done: the weights were read off the network in a fixed order (from left to right and from top to bottom) and placed in a list. Notice that each "gene" in the chromosome is a real number rather than a bit. To calculate the fitness of a given chromosome, the weights in the chromosome were assigned to the links in the corresponding network, the network was run on the training set (here 236 examples from the database), and the sum of the squares of the errors was returned. Here, an "error" was the difference between the desired output value and the actual output value. Low error meant high fitness in this case.

An initial population of 50 weights vectors was chosen randomly, with each weight being between −1.0 and +1.0. Montana and Davis tried a number of different genetic operators in various experiments. The mutation and crossover operators they used for their comparison of the genetic algorithm with backpropagation are illustrated in Figs. 7.13 and 7.14.

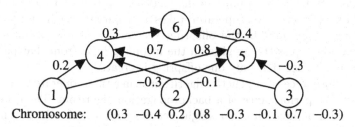

Fig. 7.12. Encoding of network weights for the genetic algorithm

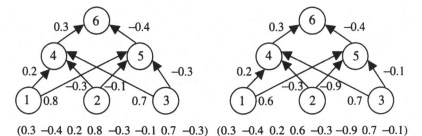

(0.3 −0.4 0.2 0.8 −0.3 −0.1 0.7 −0.3)   (0.3 −0.4 0.2 0.6 −0.3 −0.9 0.7 −0.1)

Fig. 7.13. Illustration of the mutation method. The weights on incoming links to unit 5 are mutated

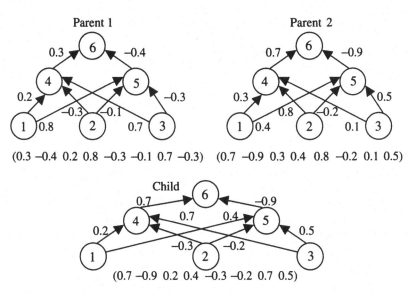

Fig. 7.14. Illustration of the crossover method. In the child network shown here, the incoming links to unit 4 come from parent 1 and the incoming links 5 and 6 come from parent 2

The mutation operator selects n non-input units, and for each incoming link to those units, adds a random value between −1.0 and +1.0 to the weight on the link. The crossover operator takes two parent weight vectors, and for each non-input unit in the offspring vector, selects one of the parents at random and copies the weights on the incoming links from that parent to the offspring. Notice that only one offspring is created.

The performance of a genetic algorithm using these operators was compared with the performance of a backpropagation algorithm. The genetic algorithm had a population of 50 weight vectors, and a rank selection method was used. The genetic algorithm was allowed to run for 200 generations. The backpropagation algorithm was allowed to run for 5000 iterations, where one iteration is a complete epoch (a complete pass through the training data). Montana and Davis found that the genetic algorithm significantly outperforms backpropagation on this task, obtaining better weight vectors more quickly.

This experiment shows that in some situations the genetic algorithm is a better training method for neural networks than simple backpropagation. This does not mean that the genetic algorithm will outperform backpropagation in all cases. It is also possible that enhancements of backpropagation might help it overcome some of the problems that prevented it from performing as well as the genetic algorithm in this experiment.

Evolving Network Architectures

Neural network researchers know all too well that the particular architecture chosen can determine the success or failure of the application, so they would like very much to be able to automatically optimize the procedure of designing an architecture for a particular application. Many believe that genetic algorithms are well suited for this task (Mitchell, 1996). There have been several efforts along these lines, most of which fall into one of two categories: direct encoding and grammatical encoding. Under direct encoding a network architecture is directly encoded into a genetic algorithm chromosome. Under grammatical encoding, the genetic algorithm does not evolve network architectures; rather, it evolves grammars that can be used to develop network architectures.

Direct Encoding

The method of direct encoding is illustrated in work done by Geoffrey Miller, Peter Todd, and Shailesh Hedge (1989), who restricted their initial project to feedforward networks with a fixed number of units for which the genetic algorithm was used to evolve the connection topology. As is shown in Fig. 7.15, the connection topology was represented by a $N \times N$ matrix (5×5 in Fig. 7.15) in which each entry encodes the type of connection from the "from unit" to the "to unit". The entries in the connectivity matrix were either "0" (meaning no connection) or "L" (meaning a "learnable" connection). Figure 7.15 also shows how the connectivity matrix was transformed into a chromosome for the genetic algorithm ("0" corresponds to 0 and "L" to 1) and how the bit string was decoded into a network. Connections that were specified to be learnable were initialized with small random weights.

Miller, Todd, and Hedge used a simple fitness-proportionate selection method and mutation (bits in the string were flipped with some low probability). Their crossover operator randomly chose a row index and swapped the corresponding rows between the two parents to create two offspring. The intuition behind that operator was similar to that behind Montana and Davis's crossover operator-each row represented all the incoming connections to a single unit, and this set was thought to be a functional building block of the

From unit	1	2	3	4	5
To unit 1	0	0	0	0	0
2	0	0	0	0	0
3	L	L	0	0	0
4	L	L	0	0	0
5	0	0	L	L	0

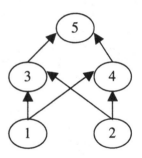

Chromosome: 0 0 0 0 0 0 0 0 0 0 1 1 0 0 0 1 1 0 0 0 0 0 1 1 0

Fig. 7.15. Illustration of Miller, Todd, and Hedge's representation scheme

network. The fitness of a chromosome was calculated in the same way as in Montana and Davis's project: for a given problem, the network was trained on a training set for a certain number of epochs, using backpropagation to modify the weights. The fitness of the chromosome was the sum of the squares of the errors on the training set at the last epoch. Again, low error translated to high fitness. Miller, Todd, and Hedge tried their genetic algorithm on several problems with very good results. The problems were relatively easy for multi-layer neural networks to learn to solve under backpropagation. The networks had different number of units for different tasks; the goal was to see if the genetic algorithm could discover a good connection topology for each task. For each run the population size was 50, the crossover rate was 0.6, and the mutation rate was 0.005. In all cases, the genetic algorithm was easily able to find networks that readily learned to map inputs to outputs over the training set with little error. However, the tasks were too easy to be a rigorous test of this method-it remains to be seen if this method can scale up to more complex tasks that require much larger networks with many more interconnections.

Grammatical Encoding

The method of grammatical encoding can be illustrated by the work of Hiroaki Kitano (1990), who points out that direct encoding approaches become increasingly difficult to use as the size of the desired network increases. As the network's size grows, the size of the required chromosome increases quickly, which leads to problems both in performance and in efficiency. In addition, since direct encoding methods explicitly represent each connection in the network, repeated or nested structures cannot be represented efficiently, even though these are common for some problems.

The solution pursued by Kitano and others is to encode networks as grammars; the genetic algorithm evolves the grammars, but the fitness is tested only after a "development" step in which a network develops from the grammar. A grammar is a set of rules that can be applied to produce a set of structures (e.g., sentences in a natural language, programs in a computer language, neural network architectures).

Kitano applied this general idea to the development of neural networks using a type of grammar called a "graph-generation grammar", a simple example of which is given in Fig. 7.16(a). Here the right-hand side of each rule is a 2×2 matrix rather than a one-dimensional string. Each lower-case letter from a through p represents one of the 16 possible 2×2 arrays of ones and zeros. There is only one structure that can be formed from this grammar: the 8×8 matrix shown in Fig. 7.16(b). This matrix can be interpreted as a connection matrix for a neural network: a 1 in row i and column i means that unit i is present in the network and a 1 in row i and column, $i \neq j$, means that there is connection from unit i to unit j. The result is the network shown in Fig. 7.16(c) which, with appropriate weights, computes the Boolean function XOR.

Kitano's goal was to have a genetic algorithm evolve such grammars. Figure 7.17 illustrates a chromosome encoding the grammar given in Fig. 7.16(a). The chromosome is divided up into separate rules, each of which consists of five elements. The first element is the left-hand side of the rule; the second through fifth elements are the four symbols in the matrix on the right-hand side of the rule. The possible values for each element are the symbols $A - Z$ and $a - p$. The first element of the chromosome is fixed to be the start symbol, S; at least one rule taking S into a 2×2 matrix is necessary to get started in building a network from a grammar.

The fitness of a grammar was calculated by constructing a network from the grammar, using backpropagation with a set of training inputs to train the resulting network to perform a simple task, and then, after training, measuring the sum of the squares of the errors made by the network on either the training set or a separate test set. The genetic algorithm used fitness-proportionate selection, multi-point crossover, and mutation. A mutation consisted of replacing one symbol in the chromosome with a randomly chosen symbol from the $A - Z$ and $a - p$ alphabets. Kitano used what he called "adaptive mutation": the probability of mutation of an offspring depended on the Hamming distance (number of mismatches) between the two parents. High distance resulted in low mutation, and vice versa. In this way, the genetic algorithm tended to respond to loss of diversity in the population by selectively raising the mutation rate.

Kitano (1990) performed a series of experiments on evolving networks for simple "encoder/decoder" problems to compare the grammatical and direct encoding approaches. He found that, on these relatively simple problems, the performance of a genetic algorithm using the grammatical encoding method consistently surpassed that of a genetic algorithm using the direct encoding method, both in the correctness of the resulting neural networks and in the speed with which they were found by the genetic algorithm. In the

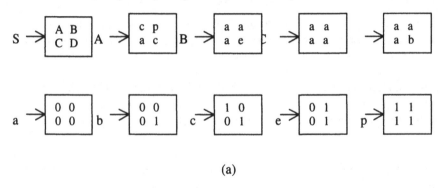

(a)

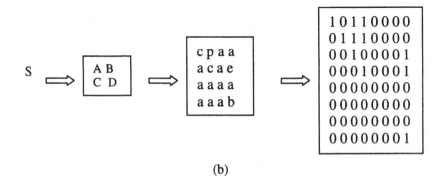

(b)

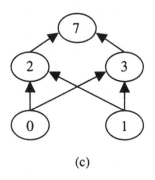

(c)

Fig. 7.16. Illustration of Kitano's graph generation grammar for the XOR problem. (a) Grammatical rules. (b) A connection matrix is produced from the grammar. (c) The resulting network

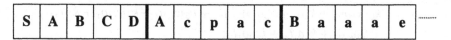

Fig. 7.17. Illustration of a chromosome encoding a grammar

grammatical encoding runs, the genetic algorithm found networks with lower error rate, and found the best networks more quickly, than in direct encoding runs. Kitano also discovered that the performance of the genetic algorithm scaled much better with network size when grammatical encoding was used.

What accounts for the grammatical encoding method's apparent superiority? Kitano argues that the grammatical encoding method can easily create "regular", repeated patterns of connectivity, and that this is a result of the repeated patterns that naturally come from repeatedly applying grammatical rules. We would expect grammatical encoding approaches, to perform well on problems requiring this kind of regularity. Grammatical encoding also has the advantage of requiring shorter chromosomes, since the genetic algorithm works on the instructions for building the network (the grammar) rather that on the network structure itself. For complex networks, the later could be huge and intractable for any search algorithm.

Evolving a Learning Rule

David Chalmers (1990) took the idea of applying genetic algorithms to neural networks in a different direction: he used genetic algorithms to evolve a good learning rule for neural networks. Chalmers limited his initial study to fully connected feedforward networks with input and output layers only, no hidden layers. In general a learning rule is used during the training procedure for modifying network weights in response to the network's performance on the training data. At each training cycle, one training pair is given to the network, which then produces an output. At this point the learning rule is invoked to modify weights. A learning rule for a single layer, fully connected feedforward network might use the following local information for a given training cycle to modify the weight on the link from input unit i to output unit j:

a_i: the activation of input i
o_j: the activation of output unit j
t_j: the training signal on output unit j
w_{ij}: the current weight on the link from i to j
    The change to make in the weight w_{ij} is a function of these values:
$\Delta w_{ij} = f(a_i, o_j, t_j, w_{ij})$.

The chromosomes in the genetic algorithm population encoded such functions.
    Chalmers made the assumption that the learning rule should be a linear function of these variables and all their pairwise products. That is, the general form of the learning rule was

$$\Delta w_{ij} = k_0(k_1 w_{ij} + k_2 a_i + k_3 o_j + k_4 t_j + k_5 w_{ij} a_i + k_6 w_{ij} o_j + k_7 w_{ij} t_j$$
$$+ k_8 a_i o_j + k_9 a_i t_j + k_{10} o_j t_j) \, .$$

The $k_m(1 < m < 10)$ are constant coefficients, and k_0 is a scale parameter that affects how much the weights can change on any cycle. Chalmer's assumption about the form of the learning rule came in part from the fact that a known good learning rule for such networks is the "delta rule". One goal of Chalmer's work was to see if the genetic algorithm could evolve a rule that performs as well as the delta rule.

The task of the genetic algorithm was to evolve values for the km's. The chromosome encoding for the set of km's is illustrated in Fig. 7.18. The scale parameter k_0 is encoded as five bits, with the zeroth bit encoding the sign (1 encoding + and 0 encoding −) and the first through fourth bits encoding an integer n : $k_0 = 0$ if $n = 0$; otherwise $|k_0| = 2^{n-9}$. Thus k_0 can take on the values $0, + -1/256, + -1/128, \ldots, + -32, + -64$. The other coefficients km are encoded by three bits each, with the zeroth bit encoding the sign and the first and second bits encoding an integer n. For $i = 1, \ldots, 10, km = 0$ if $n = 0$; otherwise $|k_m| = 2^{n-1}$.

The fitness of each chromosome (learning rule) was determined as follows. A subset of 20 mappings was selected from the full set of 30 linear separable mappings. For each mapping, 12 training examples were selected. For each of these mappings, a network was created with the appropriate number of input

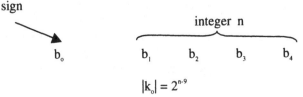

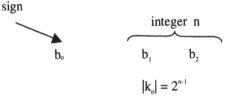

Fig. 7.18. Illustration of the method for encoding the km's

units for the given mapping. The network's weights were initialized randomly. The network was run on the training set for some number of epochs (typically 10), using the learning rule specified by the chromosome. The performance of the learning rule on a given mapping was a function of the network's error on the training set, with low error meaning high performance. The overall fitness of the learning rule was a function of the average error of 20 networks over the chosen subset of 20 mappings. This fitness was then transformed to be a percentage, where a high percentage meant high fitness.

Using this fitness measure, the genetic algorithm was run on a population of 40 learning rules, with two-point crossover and standard mutation. The crossover rate was 0.8 and the mutation rate was 0.01. Typically, over 1000 generations, the fitness of the best learning rules in the population rose from between 40% and 60% in the initial generation to between 80% and 98%. The fitness of the delta rule is around 98%, and on one out of a total of ten run the genetic algorithm discovered this rule. On three of the ten runs, the genetic algorithm discovered slight variations of this rule with lower fitness.

These results show that, given a somewhat constrained representation, the genetic algorithm was able to evolve a successful learning rule for simple single layer networks. The extent to which this method can find learning rules for more complex networks remains an open question, but these results are a first step in that direction. Chalmers suggested that it is unlikely that evolutionary methods will discover learning methods that are more powerful than backpropagation, but he speculated that genetic algorithms might be a powerful method for discovering learning rules for unsupervised learning paradigms or for new classes of network architectures.

Example of Architecture Optimization
for a Modular Neural Network

We show the feasibility of the proposed approach of using genetic algorithms for architecture optimization with a problem of system identification. We consider the non-linear function given by the following equation:

$$y = 2e^{(-0.2X)}|\sin(x)| \qquad (7.11)$$

We will use a modular neural network for system identification, as in Chap. 6, but now the genetic algorithm will automatically optimize the number of neurons in each of the modules. We will again use as range for the function the [0, 9.45] interval with a 0.01 step size, which gives us 945 points. The data is divided in the three modules as follows:

Module 1: from 0 to 3.15
Module 2: from 3.15 to 6.30
Module 3: from 6.30 to 9.45

The idea behind this data partition is that learning will be easier in each of the modules, i.e. a simple NN can learn more easily the behavior of the

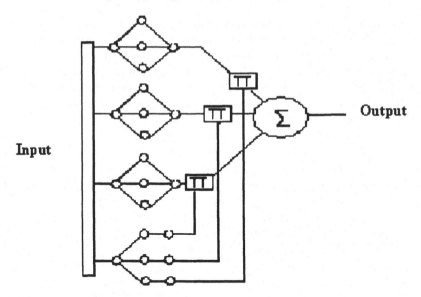

Fig. 7.19. General architecture of the modular neural network

function in one of the regions. We used three-layer feed-forward NNs for each of the modules with the Levenberg-Marquardt training algorithm. We show in Fig. 7.19 the general architecture of the MNN used in this paper.

Regarding the genetic algorithm for MNN evolution, we used a hierarchical chromosome for representing the relevant information of the network. First, we have the bits for representing the number of layers of each of the three modules and then we have the bits for representing the number of nodes of each layer. This scheme basically gives us the chromosome shown in Fig. 7.20.

CM1	M2	CM3	NC1M1	NC2M1	NC3M1	NC1M2	NC2M2	NC3M2	NC1M3	NC1M3	NC1M3
3BITS	3BITS	3BITS	8BITS	8BITS	8BITS	8BITS	8BITS	8BITS	8BITS	8BITS	8BITS

Fig. 7.20. Basic structure of the chromosome containing the information of the MNN

The fitness function used in this work combines the information of the error objective and also the information about the number of nodes as a second objective. This is shown in the following equation.

$$f(z) = \left(\frac{1}{\alpha * \text{Ranking}(ObjV1) + \beta * ObjV2} \right) * 10 \qquad (7.12)$$

The first objective is basically the average sum of squared of errors as calculated by the predicted outputs of the MNN compared with real values of the function. This is given by the following equation.

$$f_1 = \frac{1}{N} \sum_{i=1}^{N} (Y_i - y_i) \tag{7.13}$$

The parameters of the genetic algorithm are as follows:

Type of crossover operator: Two-point crossover
Crossover rate: 0.8
Type of mutation operator: Binary mutation
Mutation rate: 0.05
Population size per generation: 10
Total number of generations: 100

We show in Fig. 7.21 the topology of the final evolved modules of the neural network for the problem of function identification. As we can appreciate, from this figure, module 2 is the smallest one and module 3 is the largest one. The result of MNN evolution is a particular architecture with different size of the modules (neural networks).

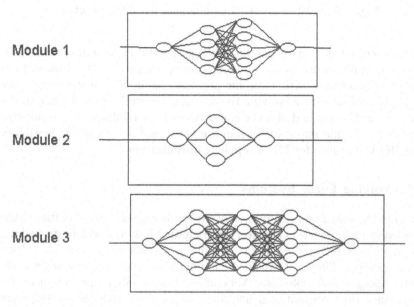

Fig. 7.21. Topology of the final evolved MNN for function identification

The MNN architecture shown in Fig. 7.21 is the best one for the specific problem of system identification. It is worthwhile noting that this network topology is difficult to design manually, for this reason the HGA approach is a good choice for neural network design and optimization. Finally, we show in Fig. 7.22 the evolution of the HGA for MNN topology optimization. From this figure, we can notice that the evolutionary approach is achieving the goal of MNN design.

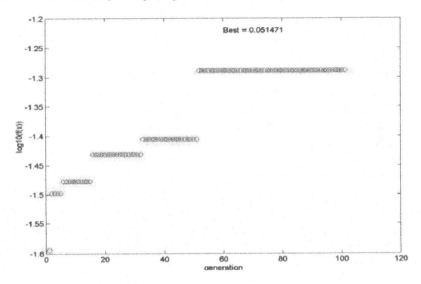

Fig. 7.22. Plot of the HGA performance for MNN evolution

We described in section our hierarchical genetic algorithm approach for modular neural network topology design and optimization. The proposed approach was illustrated with a specific problem of function identification. The best MNN is obtained by evolving the modules (single NNs) according to the error of identification and also the complexity of the modules. The results for the problem of function identification are very good and show the feasibility of the HGA approach for MNN topology optimization.

7.3.2 Evolving Fuzzy Systems

Ever since the very first introduction of the fundamental concept of fuzzy logic by Zadeh in 1973, its use in engineering disciplines has been widely studied. Its main attraction undoubtedly lies in the unique characteristics that fuzzy logic systems possess. They are capable of handling complex, non-linear dynamic systems using simple solutions. Very often, fuzzy systems provide a better performance than conventional non-fuzzy approaches with less development cost.

However, to obtain an optimal set of fuzzy membership functions and rules is not an easy task. It requires time, experience, and skills of the operator for the tedious fuzzy tuning exercise. In principle, there is no general rule or method for the fuzzy logic set-up. Recently, many researchers have considered a number of intelligent techniques for the task of tuning the fuzzy set.

Here, another innovative scheme is described (Man, Tang & Kwong, 1999). This approach has the ability to reach an optimal set of membership functions and rules without a known overall fuzzy set topology. The conceptual idea of

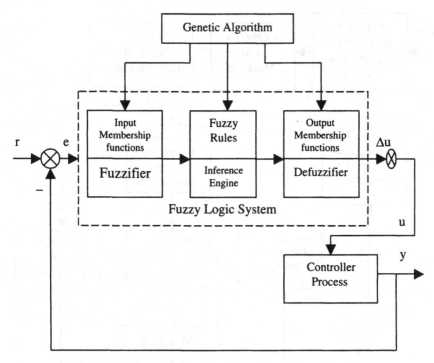

Fig. 7.23. Genetic algorithm for a fuzzy control system

this approach is to have an automatic and intelligent scheme to tune the membership functions and rules, in which the conventional closed loop fuzzy control strategy remains unchanged, as indicated in Fig. 7.23.

In this case, the chromosome of a particular is shown in Fig. 7.24. The chromosome consists of two types of genes, the control genes and parameter genes. The control genes, in the form of bits, determine the membership function activation, whereas the parameter genes are in the form of real numbers to represent the membership functions.

To obtain a complete design for the fuzzy control system, an appropriate set of fuzzy rules is required to ensure system performance. At this point it should be stressed that the introduction of the control genes is done to govern the number of fuzzy subsets in the system.

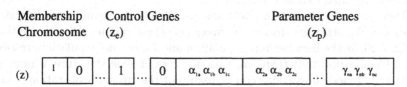

Fig. 7.24. Chromosome structure for the fuzzy system

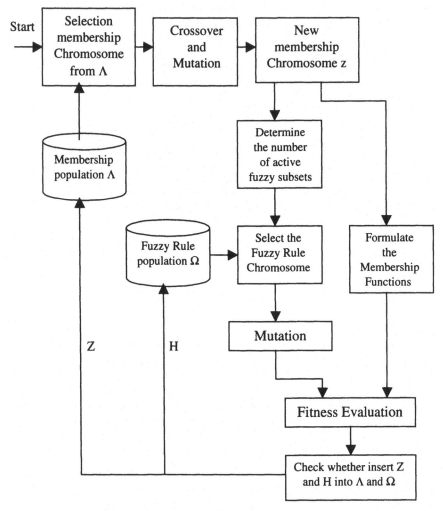

Fig. 7.25. Genetic cycle for fuzzy system optimization

Once the formulation of the chromosome has been set for the fuzzy membership functions and rules, the genetic operation cycle can be performed. This cycle of operation for the fuzzy control system optimization using a genetic algorithm is illustrated in Fig. 7.25.

There are two population pools, one for storing the membership chromosomes and the other for storing the fuzzy rule chromosomes. We can see this, in Fig. 7.26, as the membership population and fuzzy rule population, respectively. Considering that there are various types of gene structure, a number of different genetic operations can be used. For the crossover operation, a one point crossover is applied separately for both the control and parameter genes of the membership chromosomes within certain operation rates. There is no

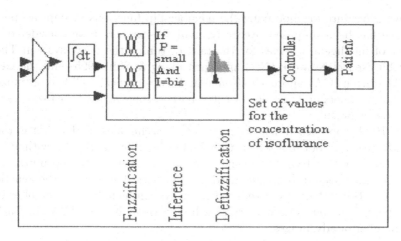

Fig. 7.26. Architecture of the fuzzy control system

crossover operation for fuzzy rule chromosomes since only one suitable rule set can be assisted.

Bit mutation is applied for the control genes of the membership chromosome. Each bit of the control gene is flipped if a probability test is satisfied (a randomly generated number is smaller than a predefined rate). As for the parameter genes, which are real number represented, random mutation is applied.

The complete genetic cycle continues until some termination criteria, for example, meeting the design specification or number of generation reaching a predefined value, are fulfilled.

Application to the Optimization of a Fuzzy Controller

We describe in this section the application of a Hierarchical Genetic Algorithm (HGA) for fuzzy system optimization (Man et al., 1999). In particular, we consider the problem of finding the optimal set of rules and membership functions for a specific application (Yen & Langari, 1999). The HGA is used to search for this optimal set of rules and membership functions, according to the data about the problem. We consider, as an illustration, the case of a fuzzy system for intelligent control.

Fuzzy systems are capable of handling complex, non-linear and sometimes mathematically intangible dynamic systems using simple solutions (Jang et al., 1997). Very often, fuzzy systems may provide a better performance than conventional non-fuzzy approaches with less development cost (Procyk & Mamdani, 1979). However, to obtain an optimal set of fuzzy membership functions and rules is not an easy task. It requires time, experience and skills of the designer for the tedious fuzzy tuning exercise. In principle, there is no general rule or method for the fuzzy logic set-up, although a heuristic and iterative procedure for modifying the membership functions to improve performance has been proposed. Recently, many researchers have considered a number of intelligent schemes for the task of tuning the fuzzy system. The noticeable Neural Network (NN) approach (Jang & Sun, 1995) and the Genetic Algorithm (GA) approach (Homaifar & McCormick, 1995) to optimize either the membership functions or rules, have become a trend for fuzzy logic system development.

The HGA approach differs from the other techniques in that it has the ability to reach an optimal set of membership functions and rules without a known fuzzy system topology (Tang et al., 1998). During the optimization phase, the membership functions need not be fixed. Throughout the genetic operations (Holland, 1975), a reduced fuzzy system including the number of membership functions and fuzzy rules will be generated. The HGA approach has a number of advantages:

(1) An optimal and the least number of membership functions and rules are obtained
(2) No pre-fixed fuzzy structure is necessary, and
(3) Simpler implementing procedures and less cost are involved.

We consider in this section the case of automatic anesthesia control in human patients for testing the optimized fuzzy controller. We did have, as a reference, the best fuzzy controller that was developed for the automatic anesthesia control (Karr & Gentry, 1993), and we consider the optimization of this controller using the HGA approach (Castillo & Melin, 2003). After applying the genetic algorithm the number of fuzzy rules was reduced from 12 to 9 with a similar performance of the fuzzy controller (Lozano, 2004). Of course, the parameters of the membership functions were also tuned by the genetic algorithm. We did compare the simulation results of the optimized fuzzy controllers obtained with the HGA against the best fuzzy controller that was obtained previously with expert knowledge, and control is achieved in a similar fashion.

The fitness function in this case can be defined in this case as follows:

$$f_i = \Sigma |y(k) - r(k)| \tag{7.14}$$

where Σ indicates the sum for all the data points in the training set, and $y(k)$ represents the real output of the fuzzy system and $r(k)$ is the reference output. This fitness value measures how well the fuzzy system is approximating the real data of the problem.

We consider the case of controlling the anesthesia given to a patient as the problem for finding the optimal fuzzy system for control (Lozano, 2004). The complete implementation was done in the MATLAB programming language. The fuzzy systems were build automatically by using the Fuzzy Logic Toolbox, and genetic algorithm was coded directly in the MATLAB language. The fuzzy systems for control are the individuals used in the genetic algorithm, and these are evaluated by comparing them to the ideal control given by the experts.

In other words, we compare the performance of the fuzzy systems that are generated by the genetic algorithm, against the ideal control system given by the experts in this application. We give more details below.

The main task of the anesthesist, during and operation, is to control anesthesia concentration. In any case, anesthesia concentration can't be measured directly. For this reason, the anesthesist uses indirect information, like the heartbeat, pressure, and motor activity. The anesthesia concentration is controlled using a medicine, which can be given by a shot or by a mix of gases. We consider here the use of isoflurance, which is usually given in a concentration of 0 to 2% with oxygen. In Fig. 7.26 we show a block diagram of the controller.

The air that is exhaled by the patient contains a specific concentration of isoflurance, and it is re-circulated to the patient. As consequence, we can measure isoflurance concentration on the inhaled and exhaled air by the patient, to estimate isoflurance concentration on the patient's blood. From the control engineering point of view, the task by the anesthesist is to maintain anesthesia concentration between the high level W (threshold to wake up) and the low level E (threshold to success). These levels are difficult to be determined in a changing environment and also are dependent on the patient's condition. For this reason, it is important to automate this anesthesia control, to perform this task more efficiently and accurately, and also to free the anesthesist from this time consuming job. The anesthesist can then concentrate in doing other task during operation of a patient.

The first automated system for anesthesia control was developed using a PID controller in the 60's. However, this system was not very successful due to the non-linear nature of the problem of anesthesia control. After this first attempt, adaptive control was proposed to automate anesthesia control, but robustness was the problem in this case. For these reasons, fuzzy logic was proposed for solving this problem. An additional advantage of fuzzy control is that we can use in the rules the same vocabulary as the medical Doctors use. The fuzzy control system can also be easily interpreted by the anesthesists.

The fuzzy system is defined as follows:

(1) Input variables: Blood pressure and Error
(2) Output variable: Isoflurance concentration
(3) Nine fuzzy if-then rules of the optimized system, which is the base for comparison
(4) 12 fuzzy if-then rules of an initial system to begin the optimization cycle of the genetic algorithm.

The linguistic values used in the fuzzy rules are the following: PB = Positive Big, PS = Positive Small, ZERO = zero, NB =Negative Big, NS = Negative Small

We show below a sample set of fuzzy rules that are used in the fuzzy inference system that is represented in the genetic algorithm for optimization.

if Blood pressure is NB and error is NB
   then conc_isoflurance is PS
if Blood pressures is PS
   then conc_isoflurance is NS
if Blood pressure is NB
   then conc_isoflurance is PB
if Blood pressure is PB
   then conc_isoflurance is NB
if Blood pressure is ZERO and error is ZERO
   then conc_isoflurance is ZERO
if Blood pressure is ZERO and error is PS
   then conc_isoflurance is NS
if Blood pressure is ZERO and error is NS
   then conc_isoflurance is PS
if error is NB
   then conc_isoflurance is PB
if error is PB
   then conc_isoflurance is NB
if error is PS
   then conc_isoflurance is NS
if Blood pressure is NS and error is ZERO
   then conc_isoflurance is NB
if Blood pressure is PS and error is ZERO
   then conc_isoflurance is PS.

The general characteristics of the genetic algorithm that are the following:

NIND = 40; % Number of individuals in each subpopulation.

MAXGEN = 300; % Maximum number of generations allowed.

GGAP = .6; %"Generational gap", which is the percentage from the complete population of new individuals generated in each generation.

PRECI = 120; % Precision of binary representations.

SelCh = select("rws", Chrom, FitnV, GGAP); % Roulette wheel method for selecting the indivuals participating in the genetic operations.

SelCh = recombin("xovmp", SelCh, 0.7); % Multi-point crossover as recombination method for the selected individuals.

ObjV = FuncionObjDifuso120_555(Chrom, sdifuso); Objective function is given by the error between the performance of the ideal control system given by the experts and the fuzzy control system given by the genetic algorithm.

In Table 7.1 we show the chromosome representation, which has 120 binary positions. These positions are divided in two parts, the first one indicates the number of rules of the fuzzy inference system, and the second one is divided again into fuzzy rules to indicate which membership functions are active or inactive for the corresponding rule.

Table 7.1. Binary Chromosome Representation

Bit assigned	Representation
1 a 12	Which rule is active or inactive
13 a 21	Membership functions active or inactive of rule 1
22 a 30	Membership functions active or inactive of rule 2
...	Membership functions active or inactive of rule...
112 a 120	Membership functions active or inactive of rule 12

We now describe the simulation results that were achieved using the hierarchical genetic algorithm for the optimization of the fuzzy control system, for the case of anesthesia control. The genetic algorithm is able to evolve the topology of the fuzzy system for the particular application. We used 300 generations of 40 individuals each to achieve the minimum error. The value of the minimum error achieved with this particular fuzzy logic controller was of 0.0064064, which is considered a small number in this application.

In Fig. 7.27 we show the simulation results of the fuzzy logic controller produced by the genetic algorithm after evolution. We used a sinusoidal input signal with unit amplitude and a frequency of 2 radians/second, with a transfer function of $[1/(0.5s+1)]$. In this figure we can appreciate the comparison of the outputs of both the ideal controller (1) and the fuzzy controller optimized by the genetic algorithm (2). From this figure it is clear that both controllers are very similar and as a consequence we can conclude that the genetic algorithm was able to optimize the performance of the fuzzy logic controller.

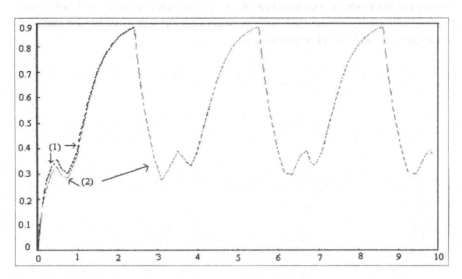

Fig. 7.27. Comparison between outputs of the ideal controller (1) and the fuzzy controller produced with the HGA (2)

We consider in this section the case of automatic anesthesia control in human patients for testing the optimized fuzzy controller. We did have, as a reference, the best fuzzy controller that was developed for the automatic anesthesia control, and we consider the optimization of this controller using the HGA approach. After applying the genetic algorithm the number of fuzzy rules was reduced from 12 to 9 with a similar performance of the fuzzy controller. Of course, the parameters of the membership functions were also tuned by the genetic algorithm. We did compare the simulation results of the optimized fuzzy controllers obtained with the HGA against the best fuzzy controller that was obtained previously with expert knowledge, and control is achieved in a similar fashion. Since simulation results are similar, and the number of rules was reduced, we can conclude that the HGA approach is a good alternative for designing fuzzy systems.

7.4 Summary

We have presented in this chapter the basic concepts of genetic algorithms and their applications. These optimization methodologies are motivated by nature's wisdom. Genetic algorithms emulate the process of evolution in nature. We have presented classical examples of the application of these optimization techniques. We have also presented the application of genetic algorithms to the problems of optimizing neural networks and fuzzy systems. Genetic algorithms can then be viewed as a technique for efficient design of intelligent systems, because they can be used to optimize the weights or architecture of the neural network, or the number of rules in a fuzzy system. In later chapters we will make use of this fact to design intelligent systems for pattern recognition in real world applications.

8

Clustering with Intelligent Techniques

Cluster analysis is a technique for grouping data and finding structures in data. The most common application of clustering methods is to partition a data set into clusters or classes, where similar data are assigned to the same cluster whereas dissimilar data should belong to different clusters. In real-world applications there is very often no clear boundary between clusters so that fuzzy clustering is often a good alternative to use. Membership degrees between zero and one are used in fuzzy clustering instead of crisp assignments of the data to clusters.

Pattern recognition techniques can be classified into two broad categories: *unsupervised* techniques and *supervised* techniques. An unsupervised technique does not use a given set of unclassified data points, whereas a supervised technique uses a data set with known classifications. These two types of techniques are complementary. For example, unsupervised clustering can be used to produce classification information needed by a supervised pattern recognition technique. In this chapter, we first give the basics of unsupervised clustering. The Fuzzy C-Means algorithm (FCM), which is the best known unsupervised fuzzy clustering algorithm is then described in detail. Supervised pattern recognition using fuzzy logic will also be mentioned. Finally, we will describe the use of neural networks for unsupervised clustering and hybrid approaches.

8.1 Unsupervised Clustering

Unsupervised clustering is motivated by the need of finding interesting patterns or groupings in a given data set. For example, in voting analysis one may want to collect data about a group of voters (e.g. through a survey or interviews) and analyze these data to find interesting groupings of voters. The result of such an analysis can be used to plan the strategies of a particular candidate in an election.

Patricia Melin and Oscar Castillo: *Hybrid Intelligent Systems for Pattern Recognition Using Soft Computing*, StudFuzz **172**, 169–184 (2005)
www.springerlink.com

In the area of pattern recognition and image processing, unsupervised clustering is often used to perform the task of "segmenting" the images (i.e., partitioning pixels of an image into regions that correspond to different objects or different faces of objects in the images). This is because image segmentation can be considered as a kind of data clustering problem where each datum is described by a set of image features of a pixel.

Conventional clustering algorithms find what is called a "hard partition" of a given data set based on certain criteria that evaluate the goodness of a partition. By a "hard partition" we mean that each datum belongs to exactly one cluster of the partition. We can define the concept of a "hard partition", more formally, as follows.

Definition 8.1. *Let X be a data set, and x_i be an element of X. A partition $P = \{C_1, C_2, \ldots, C_l\}$ of X is "hard" if and only if the two following conditions are satisfied:*

(1) For all $x_i \in X$, there exists a $Cj \in P$ such that $x_i \in Cj$
(2) For all $x_i \in X, x_i \in Cj \rightarrow x_i \notin Ci$ where $i \neq j, Ci, Cj \in P$.

The first condition in this definition states that the partition has to cover all data points in X, and the second condition states that all the clusters in the partition are mutually exclusive (in other words, there can not be intersection between clusters).

In many real-world clustering problems, however, some data points can partially belong to multiple clusters, rather than to a single cluster exclusively. For example, a pixel in a medical image, of the human brain of a patient, may correspond to a mixture of two different types of cells. On the other hand, in the voting analysis example, a particular voter maybe a "borderline case" between two groups of voters (e.g., between moderate conservatives and moderate liberals). These examples motivate the need for having "soft partitions" and "soft clustering algorithms".

A soft clustering algorithm finds a "soft partition" of a given data set based on certain criteria. In a soft partition, a datum can partially belong to multiple clusters. A formal definition of this concept is the following.

Definition 8.2. *Let X be a set of data, and x_i be an element of X. A partition $P = \{C_1, C_2, \ldots, C_l\}$ of X is a "soft partition" if and only if the following two conditions are satisfied:*

(1) For all $x_i \in X$, for all $C_j \in P$, then $0 \leq \mu_{Cj}(x_i) \leq 1$
(2) For all $x_i \in X$, there exists $C_j \in P$ such that $\mu_{Cj}(x_i) > 0$.

Where $\mu_{Cj}(x_i)$ denotes the degree of membership to which x_i belongs to cluster C_j.

A particular type of soft clustering is one that ensures that the membership degree of a point x in all clusters add up to one. More formally,

$$\Sigma_j \mu_{Cj}(x_i) = 1 \quad \text{For all } x_i \in X$$

A soft partition that satisfies this additional condition is called a called a constrained soft partition. The fuzzy c-means clustering algorithm produces a constrained soft partition, as we will see later in this chapter.

A constrained soft partition can also be generated by what it is called a probabilistic clustering algorithm. Even though both fuzzy c-means and probabilistic clustering produce a partition with similar properties, the clustering criteria underlying these algorithms are very different. We will concentrate in this chapter on fuzzy clustering, but this does not mean that probabilistic clustering is not a good method. The fuzzy c-means algorithm generalizes a hard clustering algorithm called the c-means algorithm. The hard c-means algorithm aims to identify compact well-separated clusters. Informally, a "compact" cluster has a "ball-like" shape. The center of the ball is called "the center" or "the prototype" of the cluster. A set of clusters are "well separated" when any two points in a cluster are closer than the shortest distance between two points in different clusters.

Assuming that a data set containsccompact, well-separated clusters, the goals of the hard c-means algorithm are the following:

(1) To find the centers of these clusters, and
(2) To determine the clusters (i.e., labels) of each point in the data set.

In fact, the second goal can easily be achieved once we accomplish the first goal, based on the assumption that clusters are compact and well separated. Given the cluster centers, a point in the data set belongs to the cluster whose center is the closest.

In order to achieve the first goal, we need to establish a criterion that can be used to search for these cluster centers. One of the criteria that are used is the sum of the distances between points in each cluster and their centers. This can stated formally as follows:

$$J(P, V) = \Sigma_j \Sigma_{xi} ||x_i - v_j||^2 \tag{8.1}$$

Where $j = 1$ to $j = c, x_i \in C_j$ and V is a vector of cluster centers to be identified. This criterion is useful because a set of true cluster centers will give a minimal value of J for a given data set. Based on these observations, the hard c-means algorithm tries to find the cluster centers that minimize the value of the objective function J. However, J is also a function of the partition P, which is determined by the cluster centers V. As a consequence, the hard c-means algorithm searches for the true cluster centers by iterating the following two steps:

(1) Calculating the current partition based on the current clusters,
(2) Modifying the current clusters using gradient descent method to minimize the objective function J.

The iterations terminate when the difference between cluster centers in two consecutive iterations is smaller than a specified threshold. This means that the clustering algorithm has converged to a local minimum of J.

8.2 Fuzzy C-Means Algorithm

The Fuzzy C-Means algorithm (FCM) generalizes the hard c-means algorithm to allow points to partially belong to multiple clusters. Therefore, it produces a soft partition for a given data set. In fact, it generates a constrained soft partition. To achieve this, the objective function of hard c-means has to be extended in two ways:

(1) the fuzzy membership degrees in clusters were incorporated into formula, and
(2) an additional parameter m was introduced as a weight exponent in the fuzzy membership.

The extended objective function, denoted J_m, is defined as follows:

$$J_m(P, V) = \Sigma_j \Sigma_{xi} \mu_{Cj}(x_k)^m ||x_i - v_j||^2 \qquad (8.2)$$

where P is a fuzzy partition of the data set X formed by $C_1, C_2, \ldots, C_k$. The parameter m is a weight that determines the degree to which partial members of a cluster affect the clustering result.

Like hard c-means, the fuzzy c-means algorithm tries to find a good partition by searching prototypes vi that minimize the objective function J_m. Unlike hard c-means, however, the fuzzy c-means algorithm also needs to search for membership functions μ_{Cj} that minimize J_m. To accomplish these two objectives, a necessary condition for the local minimum of J_m has to be obtained from J_m. Basically, the partial derivatives of J_m have to be zero and then solved simultaneously. This condition, which is formally stated below, serves as the foundation for the fuzzy c-means algorithm.

Theorem 8.1. *Fuzzy C-Means Theorem*
A constrained fuzzy partition $\{C_1, C_2, \ldots, C_k\}$ is a local minimum of the objective function J_m only if the following conditions are satisfied:

$$\mu_{Ci}(x) = 1/ \left[\Sigma_j (||x - v_j||^2/||x - v_j||^2)^{(1/m-1)} \right] \quad 1 \le i \le k, x \in X \quad (8.3)$$
$$v_i = [\Sigma_x (\mu_{Ci}(x))^m (x)]/[\Sigma_x (\mu_{Ci}(x))^m] \quad 1 \le i \le k \qquad (8.4)$$

Based on the equations of this theorem, FCM updates the prototypes and the membership functions iteratively using (8.3) and (8.4) until a termination criterion is satisfied. We can specify the FCM algorithm as follows:

Step 1: Initialize prototype $V = \{v_1, v_2, \ldots, v_c\}$.
Step 2: Make $V^{old} \leftarrow V$.
Step 3: Calculate membership functions with (8.3).
Step 4: Update the prototype v_i in V using (8.4).
Step 5: Calculate $E = \Sigma_i ||v^{old} - v_i||$
Step 6: If $E > \varepsilon$ then go to Step 2.
Step 7: If $E \le \varepsilon$ then output the final result.

In the previous algorithm, c is the number of clusters to form, m is the parameter of the objective function, and ε is a threshold value (usually very small) for the convergence criteria.

We have to say that the FCM algorithm is guaranteed to converge for $m > 1$. This important convergence theorem was established by Bezdek in 1990. FCM finds a local minimum of the objective function J_m. This is because the FCM theorem is derived from the condition that the gradient of J_m should be zero for a FCM solution, which is satisfied by all the local minimums. Finally, we have to say that the result of applying FCM to a given data set depends not only on the choice of the parameters m and c, but also on the choice of the initial prototypes.

We now consider a simple example to illustrate the concepts and ideas of the FCM algorithm. We consider a set of randomly generated points, and the application of the FCM algorithm to find four clusters. In Fig. 8.1 we can appreciate a simulation of the FCM algorithm in which the initial cluster centers (at the origin) move toward the "right" positions, which shows that the algorithm finds the optimal cluster centers.

After the clustering process stops, the final cluster centers are found. These cluster centers are the ones that minimize the objective function J_m. We show in Fig. 8.2 the final result of the FCM algorithm in which the final cluster center positions are indicated as black circles.

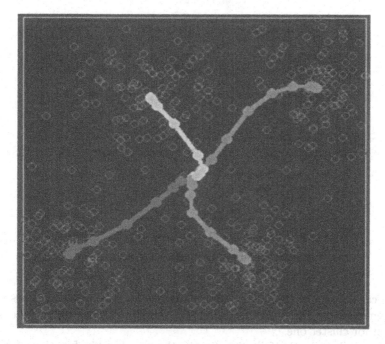

Fig. 8.1. Evolution of the FCM algorithm

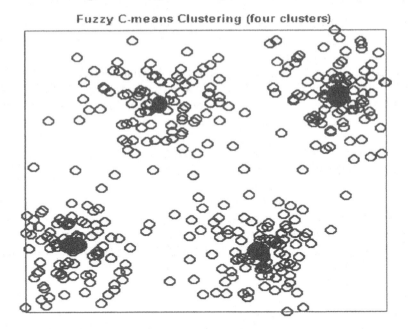

Fig. 8.2. Final cluster centers after the application of FCM

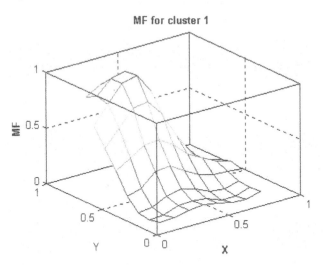

Fig. 8.3. Membership function for cluster one

We can also show the membership functions for each of the clusters that were generated by the FCM algorithm. We show in Fig. 8.3 the membership function for cluster one.

We now show in the following figures the membership functions for the other clusters in this example. In Fig. 8.4 we can appreciate the membership

MF for cluster 2

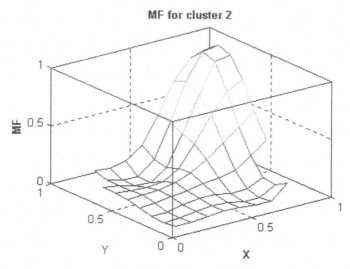

Fig. 8.4. Membership function of cluster two as a result of the FCM

MF for Cluster 3

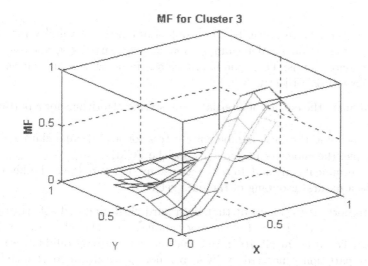

Fig. 8.5. Membership function of cluster three as a result of the FCM

function formed by FCM for cluster two. On the other hand, in Fig. 8.5, we can appreciate the membership function of cluster three. Finally, in Fig. 8.6, the membership function of cluster four is shown.

8.2.1 Clustering Validity

One of the main problems in clustering is how to evaluate the clustering result of a particular algorithm. The problem is called "clustering validity"

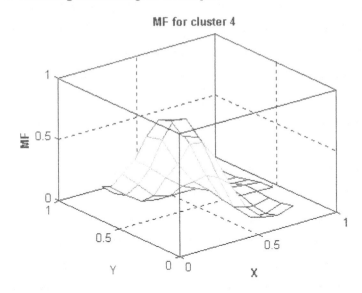

Fig. 8.6. Membership function of cluster four as a result of the FCM

in the literature. More precisely, the problem of clustering validity is to find
an objective criterion for determining how good a partition generated by a
clustering algorithm is. This type of criterion is important because it enables
us to achieve three objectives:

(1) To compare the output of alternative clustering algorithms for a particular
    data set.
(2) To determine the best number of clusters for a particular data set (for
    example, the choice of parameter c for the FCM).
(3) To determine if a given data set contains any structure (i.e., whether there
    exists a natural grouping of the data set).

It is important to point out that both hard clustering and soft clustering
need to consider the issue of clustering validity, even though their methods
may differ. We will concentrate in this chapter on describing validity measures
of a fuzzy partition generated by FCM or, more generally, a constrained soft
partition of a data set.

Validity measures of a constrained soft partition fall into three categories:
(1) membership-based validity measures, (2) geometry-based validity measures, and (3) performance-based validity measures. The membership-based
validity measures calculate certain properties of the membership functions in
a constrained soft partition. The geometry-based validity measures consider
geometrical properties of a cluster (like area or volume) as well as geometrical
relationships between clusters (for example, the separation) in a soft partition.
The performance-based validity measures evaluate a soft partition based on

its performance for a predefined goal (for example, minimum error rate for a classification task).

Jim Bezdek introduced the first validity measure of a soft partition, which is called "partition coefficient" (Yen & Langari, 1999). The partition coefficient has the goal of measuring the degree of fuzziness of clusters. The reasoning behind this validity measure is that the fuzzier the clusters are, the worse the partition is. The formula for this validity measure, denoted v_{pc}, is

$$v_{pc} = (1/n)\Sigma_i\Sigma_j\mu_{Ci}(x_j), \quad i = 1,\ldots,c, \quad j = 1,\ldots,n \,. \tag{8.5}$$

Subsequently, Bezdek also introduced another membership-based validity measure called "partition entropy", which is denoted as v_{pe} and defined formally as follows

$$v_{pe} = (-1/n)\Sigma_i\Sigma_j[\mu_{Ci}(x_j)\log_a(\mu_{Ci}(x_j))], \quad i = 1,\ldots,c, \quad j = 1,\ldots,n \,. \tag{8.6}$$

where $a \in (1,\infty)$ is the logarithmic base. The entropy measure increases as the fuzziness of the partition decreases. Therefore, a clustering achieved with lower partition entropy is preferred. The two membership-based validity measures are related in the following ways for a constrained soft partition:

(1) $v_{pc} = 1 \Leftrightarrow v_{pe} = 0 \Leftrightarrow$ the partition is hard.
(2) $v_{pc} = 1/c \Leftrightarrow v_{pe} = \log_a(c) \Leftrightarrow \mu_{Ci}(x_j) = 1/c$ for all i, j.

The two cases above correspond to the extreme situations. The first case is the least fuzzy and therefore is the one preferred (at least ideally). The second case is the fuzziest and therefore the least preferred by the validity measures.

X. Xie and G. Beni introduced a validity measure that considers both the compactness of clusters as well as the separation between clusters. The basic idea of this validity measure is that the more compact the clusters are and the further the separation between clusters, the more desirable the partition. To achieve this measure, the Xie-Beni validity index (denoted as v_{xB}) is defined formally as follows:

$$v_{xB} = [\Sigma_i\sigma_i/n]\left[1/d_{\min}^2\right] \tag{8.7}$$

where σ_i is the variation of cluster C_i defined as follows:

$$\sigma_i = \Sigma_j[\mu_{Ci}(x_j)]\|x_j - v_i\|^2 \tag{8.8}$$

n is the cardinality of the data set and $d_{\min}$ is the shortest distance between cluster centers defined as

$$d_{\min} = \min\|v_j - v_i\|, \quad i \neq j \,, \tag{8.9}$$

The first term in (8.7) is a measure of non-compactness, and the second term is a measure of non-separation. Hence, the product of the two terms reflects the degree to which the clusters in the partition are not compact and not well separated. Obviously, the lower the cluster index, the better the soft partition is.

8.2.2 Extensions to Fuzzy C-Means

A major extension to FCM is to generalize the distance measure between data x and prototype vj from the usual Euclidean distance:

$$\text{dist}(x_i, vj) = ||x_i - v_j||^2 \qquad (8.10)$$

to the generalized distance

$$\text{dist}(x_i, vj) = (x_i - v_j)^T A_j (x_i - v_j) \qquad (8.11)$$

where A_j is a symmetric $d \times d$ matrix (d is the dimensionality of x_i and v_j). This enables an extended FCM to adapt to different hyper-ellipsoidal shapes of different clusters by adjusting the matrix A_j.

D. Gustafsun and W. Kessel were the first to propose such an extension of the matrix Aj, they developed a modified FCM that dynamically adjusts $(A_1, A_2, \ldots, A_c)$ such that these matrices adapt to the different hyper-ellipsoidal shape of each cluster.

8.3 Mountain Clustering Method

The "mountain clustering method", as proposed by Yager & Filev (1994), is a relatively simple and effective approach to approximate the estimation of cluster centers on the basis of a density measure called the "mountain function". This method can be used to obtain initial cluster centers that are required by more sophisticated cluster algorithms, such as the FCM algorithm introduced in the previous section. It can also be used as a quick stand-alone method for approximate clustering. The method is based on what a human does in visually forming clusters of a data set.

The first step involves forming a grid on the data space, where the intersections of the grid lines constitute the candidates for cluster centers, denoted as a set V. A finer gridding increases the number of potential clustering centers, but it also increases the computation required. The grid is generally evenly spaced, but it not a requirement. We can have unevenly spaced grids to reflect "a priori" knowledge of data distribution. Moreover, if the data set itself (instead of the grid points) is used as the candidates for cluster centers, then we have a variant called subtractive clustering.

The second step entails building a mountain function representing a data density measure. The height of the mountain function at a point $\mathbf{v} \in V$ is equal to

$$m(\mathbf{v}) = \Sigma_i \exp[(-||\mathbf{v} - x_i||^2)/2\sigma^2] \qquad (8.12)$$

where x_i is the ith data point and σ is a constant, which is application-specific. The preceding equation, implies that each data point x_i contributes to the height of the mountain function at $\mathbf{v}$, and the contribution is inversely proportional to the distance between x_i and $\mathbf{v}$. The mountain function can be viewed as a measure of "data density", since it tends to be higher if more data points are located nearby, and lower if fewer data points are around. The constant σ determines the height as well as the smoothness of the resultant mountain function. The clustering results are normally insensitive to the value of σ, as long as the data set is of sufficient size and is well clustered.

The third step involves selecting the cluster centers by sequentially destroying the mountain function. We first find the point in the candidate centers V that has the greatest value for the mountain function; this becomes the first cluster center c_1. Obtaining the next cluster center requires eliminating the effect of the recently found center, which is typically surrounded by a number of grid points that also have high-density scores. This is realized by revising the mountain function; a new mountain function is formed by subtracting a scaled Gaussian function centered at c_1:

$$m_{\text{new}}(\mathbf{v}) = m(\mathbf{v}) - m(c_1) \exp[(-||\mathbf{v} - c_1||^2)/2\beta^2] \qquad (8.13)$$

The subtracted amount is inversely proportional to the distance between $\mathbf{v}$ and the recently found center c_1, as well as being proportional to the height $m(c_1)$ at the center. Note that after subtraction, the new mountain function $m_{\text{new}}(\mathbf{v})$ reduces to zero at $\mathbf{v} = c_1$.

After subtraction, the second cluster center is again selected as the point in V that has the largest value for the new mountain function. This process of revising the mountain function and finding the next cluster center continues until a sufficient number of cluster centers is obtained.

We now give a simple example to illustrate the ideas mentioned above. Figure 8.7(a) shows a set of two-dimensional data, in which three clusters can be very easily recognized. However, for higher dimensional data sets is very difficult to visualize the clusters, and for this reason, methods like this are very useful. In this example, the mountain method is used to find the three clusters. To show the effect of changing σ, Figs. 8.7(b) through 8.7(d) are the surface plots of the mountain functions with σ equal to 0.02, 0.1, and 0.2, respectively. Obviously, σ affects the mountain function's height as well as its smoothness; therefore, the value of σ should be selected carefully considering both the data size and input dimension.

Once the σ value is determined (0.1 in this example) and the mountain function is constructed, we begin to select clusters and revise the mountain functions sequentially. This is shown in Figs. 8.8(a), (b), (c), and (d), with β equal to 0.1 in (8.13).

8.4 Clustering of Real-World Data with Fuzzy Logic

Now we will consider a more realistic application of clustering techniques. We consider in this example clustering of World Bank data for 98 countries

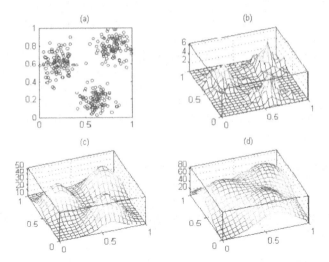

Fig. 8.7. Mountain construction: (a) original data set, (b) mountain function with $\sigma = 0.02$, (c) mountain function with $\sigma = 0.1$, and (d) mountain function with $\sigma = 0.2$

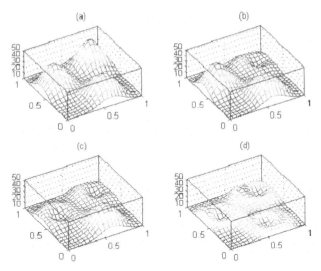

Fig. 8.8. Mountain destruction with $\beta = 0.1$: (a) original mountain function with $\sigma = 0.1$; (b) mountain function after the first reduction; (c) mountain function after the second reduction; (d) mountain function after the third reduction

with respect to two variables. The economic variables are the Electrical Power Consumption (x), and High Technology Exports (y) for 2001. The application of intelligent clustering to this data set has the goal of grouping similar countries with respect to these economic variables. The cluster will represent groups of countries with similarities in these variables. Of course, we expect that countries with low electrical power consumption will also have low high technology exports. However, other groups are more difficult to identify.

The results of the clustering will be useful to people working in Economics because a great part of their research consists in analyzing economic data and constructing models based on these studies. Of course, the clustering of countries can also be done using more variables or economic indicators that are also available in the web page of the World Bank.

We used the FCM algorithm with different number of clusters until the optimal number was found. The best results were obtained for a number $c = 6$ clusters. We show in Fig. 8.9 the final result of the FCM for the data set of the 98 countries. We also show in Fig. 8.10 the membership function obtained for cluster one.

The other membership functions for the other clusters are similar. The final clusters give the groups of countries that were grouped together by the FCM algorithm. The results were checked by the Economists and were find to be consistent with what they believe is true for these countries.

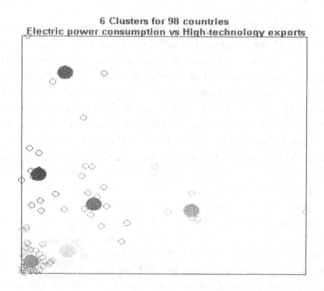

Fig. 8.9. Clustering of the data set for 98 countries with FCM

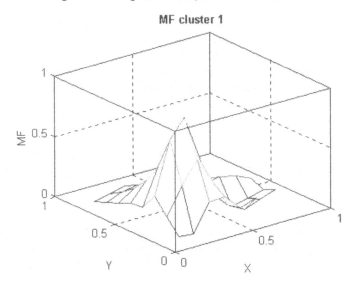

Fig. 8.10. Membership function for cluster number one of World Bank data set

8.5 Clustering of Real-World Data with Unsupervised Neural Networks

We will now consider the application of unsupervised neural networks (described in Chap. 5) for achieving intelligent clustering. This type of neural network can also be used to cluster a data set, but the main difference with respect to fuzzy clustering is that neural networks do not use membership functions. We can say that unsupervised neural network perform hard clustering on the data sets. However, neural network may have other advantages because their basis for performing clustering is based on models of learning.

We will consider again the World Bank data set of 98 countries with the two variables: electric power consumption, and high technology exports. We will apply a competitive learning neural network to this data set to form the groups of countries. These groups will form according to the similarities between the values of the countries in the data set. We show in Fig. 8.11 the results of applying a competitive neural network with 6000 epochs and six clusters.

In this case, we can appreciate that clustering has not worked well because there are points that were not considered. However, if we change the number of epochs of learning in the same neural network, we obtain the results shown in Fig. 8.12. These results are better because one of the clusters takes into account the upper points in the figure. Also, if we compare Fig. 8.12 with Fig. 8.9, we can appreciate an agreement with the results of fuzzy clustering.

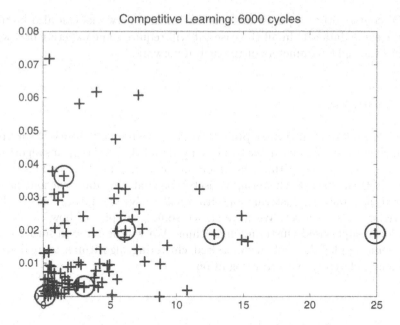

Fig. 8.11. Application of a competitive neural network for the World Bank data with 6000 epochs and a learning rate of 0.01

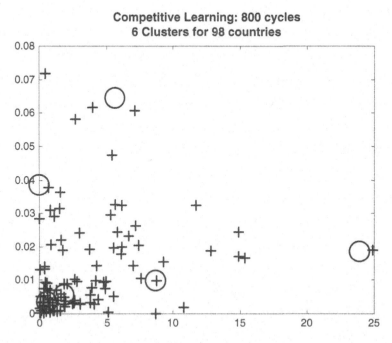

Fig. 8.12. Application of a competitive neural network for the World Bank data with 800 epochs and a learning rate of 1.3

Of course, other types of unsupervised neural networks can also be used for the same data set. In all the cases, it will require some experimental work to find the right parameters of the neural network.

8.6 Summary

We have presented in this chapter the basic concepts and theory of unsupervised clustering. We have described in some detail the hard c-means algorithm, the fuzzy c-means algorithm, the mountain clustering method, and clustering validity. Unsupervised clustering is useful for analyzing data without having desired outputs; the clustering algorithm will evolve to capture density characteristics of a data set. We have shown some examples of how to use this type of unsupervised clustering algorithms. We will describe in future chapters some applications of unsupervised clustering algorithms to real world problems related to pattern recognition.

9

Face Recognition
with Modular Neural Networks
and Fuzzy Measures

We describe in this chapter a new approach for face recognition using modular neural networks with a fuzzy logic method for response integration. We describe a new architecture for modular neural networks for achieving pattern recognition in the particular case of human faces. Also, the method for achieving response integration is based on the fuzzy Sugeno integral. Response integration is required to combine the outputs of all the modules in the modular network. We have applied the new approach for face recognition with a real database of faces from students and professors of our institution. Recognition rates with the modular approach were compared against the monolithic single neural network approach, to measure the improvement. The results of the new modular neural network approach gives excellent performance overall and also in comparison with the monolithic approach. The chapter is divided as follows: first we give a brief introduction to the problem of face recognition, second we describe the proposed architecture for achieving face recognition, third, we describe the fuzzy method for response integration, and finally we show a summary of the results and conclusions.

9.1 Introduction

Automatic face detection and recognition has been a difficult problem in the field of computer vision for several years. Although humans perform this task in an effortless manner, the underlying computations within the human visual system are of tremendous complexity. The seemingly trivial task of finding and recognizing faces is the result of millions of years of evolution and we are far from fully understanding how the human brain performs it. Furthermore, the ability to find faces visually in a scene and recognize them is critical for humans in their everyday activities. Consequently, the automation of this task would be useful for many applications including security, surveillance, affective computing, speech recognition assistance, video compression and animation.

Patricia Melin and Oscar Castillo: *Hybrid Intelligent Systems for Pattern Recognition Using Soft Computing*, StudFuzz **172**, 185–206 (2005)
www.springerlink.com

However, to this date, no complete solution has been proposed that allows the automatic recognition of faces in real images.

Robust face recognition requires the ability to recognize identity despite many variations in appearance that the face can have in a scene. The face is a three-dimensional object, which is illuminated from a variety of light sources and surrounded by arbitrary background data (including other faces). Therefore, the appearance a face has when projected onto a two-dimensional image can vary tremendously. If we want to have a system capable of performing recognition, we need to find and recognize faces despite these variations. Additionally, our detection and recognition scheme must also be capable of tolerating variations in the faces themselves. The human face is not a unique rigid object. There are billions of different faces and each them can assume a variety of deformations. Inter-personal variations can be due to deformations, expression, aging, facial hair, and cosmetics. Furthermore, the output of the recognition system has to be accurate. A recognition system has to associate an identity or name for each face it comes across by matching it to a large database of individuals. Simultaneously, the system must be robust to typical image-acquisition problems such as noise, video-camera distortion and image resolution. Thus, we are dealing with a multi-dimensional detection and recognition problem. One final constraint is the need to maintain the usability of the system on contemporary computational devices. In other words, the processing involved should be efficient with respect to run-time and storage space.

The basic idea of the new approach presented in this chapter is to divide a human face in to three different regions: the eyes, the nose and the mouth. Each of these regions is assigned to one module of the neural network. In this way, the modular neural network has three different modules, one for each of the regions of the human face. At the end, the final decision of face recognition is done by an integration module, which has to take into consideration the results of each of the modules. In our approach, the integration module uses the fuzzy Sugeno integral to combine the outputs of the three modules. The fuzzy Sugeno integral allows the integration of responses from the three modules of the eyes, nose and mouth of a human specific face. Other approaches in the literature use other types of integration modules, like voting methods, majority methods, and neural networks.

Response integration methods for modular neural networks that have been studied, to the moment, do not solve well real recognition problems with large sets of data or in other cases reduce the final output to the result of only one module. Also, in the particular case of face recognition, methods of weighted statistical average do not work well due to the nature of the face recognition problem. For these reasons, a new approach for face recognition using modular neural networks and fuzzy integration of responses was proposed in this chapter.

The new approach for face recognition was tested with a database of students and professors from our institution. This database was collected at our institution using a digital camera. The experimental results with our new approach for face recognition on this database were excellent and for this reason we think that the approach can be applied in other cases, with different databases, for applications in industrial security, airport security, and similar ones.

Face recognition has a number of advantages over some of the other biometrics measures used in practice. Whereas many biometrics require the subjects cooperation and awareness in order to perform an identification or verification, such as looking into an eye scanner or placing their hand on a fingerprint reader, face recognition could be performed even without the subject's knowledge. Secondly, the biometrics data used to perform the recognition task is in a format that is readable and understood by a human. This means that a potential face recognition system can always be backed up and verified by a human. For example, assuming that a person was incorrectly denied access to a site by an automatic recognition system, that decision could easily be corrected by a security guard that would compare the subject's face with the stored image, whereas this would not be possible with other biometrics such as iris. Other advantages are that there is no association with crime as with fingerprints (few people would object to looking at a camera) and many existing systems already store face images (such as in police databases).

The term face recognition encompasses three main procedures. The preliminary step of face detection, which may include some feature localization, is often necessary if no manual (human) intervention is to be used. Many methods have been used to accomplish this, including template based techniques, motion detection, skin tone segmentation, principal component analysis and classification by neural networks. All of which present the difficult task of characterizing "non-face" images. Also, many of the algorithms currently available are only applicable to specific situations: assumptions are made regarding the orientation and size of the face in the image, lighting conditions, background, and subject's cooperation. The next procedure is verification. This describes the process by which two face images are compared, producing a result to indicate if the two images are of the same person. Another procedure, which is sometimes more difficult, is identification. This requires a probe image, for which a matching image is searched for in a database of known people. We will concentrate in this chapter more on verification and identification.

9.2 State of the Art in Face Recognition

Research in intensity image face recognition generally falls into two categories (Chellapa, 1994): holistic (global) methods and feature-based methods. Feature-based methods rely on the identification of certain important points on the face such as the eyes, the nose, the mouth, etc. The location of those points can be determined and used to compute geometrical relationships between the points as well as to analyse the surrounding region locally. Thus, independent processing of the eyes, the nose, and other important points is performed and then combined to produce recognition of the face. Since detection of feature points precedes the analysis, such a system is robust to position variations in the image. Holistic methods treat the image data simultaneously without attempting to find individual points. The face is recognized as one entity without explicitly isolating different regions in the face. Holistic techniques use statistical analysis, neural networks and transformations. They also usually require large samples of training data. The advantage of holistic methods is that they use the face as a whole and do not destroy any information by exclusively processing only certain important points. Thus, they generally provide more accurate recognition results. However, such techniques are sensitive to variations in position, scale and so on, which restricts their use to standard, frontal mug-shot images.

Early attempts at face recognition were mostly feature-based. These include Kanade's work where a series of important points are detected using relatively simple image processing techniques (edge maps, signatures, etc.) and their Euclidean distances are then used as a feature vector to perform recognition (Kanade, 1973). More sophisticated, feature extraction, algorithms were proposed by (Yuille et al., 1989). These use deformable templates that translate, rotate and deform in search of a best fit in the image. Often, these search techniques use a knowledge-based system or heuristics to restrict the search space with geometrical constraints (i.e. the mouth must be between the eyes), like in the work by (Craw et al., 1992). Unfortunately, such energy minimization methods are extremely computationally expensive and can get trapped in local minima. Furthermore, a certain tolerance must be given to the models since they can never perfectly fit the structures in the image. However, the use of a large tolerance value tends to destroy the precision required to recognize individuals on the basis of the model's final best-fit parameters. Nixon proposes the use of Hough transform techniques to detect structures more efficiently (Nixon, 1995). However, the problem remains that these detection-based algorithms need to be tolerant and robust and this often makes them insensitive to the minute variations needed for recognition. Recent research in geometrical, feature-based recognition (Cox et al., 1995) reported 95% recognition. However, the 30 features points used for each face were manually extracted from each image. Had some form of automatic localization been used, it would have generated poorer results due to lower precision. In fact, even the most precise deformable template matching algorithms such as (Roeder and Li, 1995) and feature detectors generally have significant errors in detection (Colombo et al., 1995). This is also true for other feature detection schemes such as Reisfeld's symmetry operator (Reisfeld, 1994) and Graf's filtering and morphological operations (Graf et al., 1995). Essentially, current systems for automatic detection of important points are not accurate enough to obtain high recognition rates exclusively on the basis of simple geometrical statistics of the localization.

Holistic techniques have recently been popularized and generally involve the use of transforms to make the recognition robust to slight variations in the image. Rao developed an iconic representation of faces by transforming them into a linear combination of natural basis functions (Rao and Ballard, 1995). Manjunath uses a wavelet transform to simultaneously extract feature points and to perform recognition on the basis of their Gabor wavelet jets (Manjunath et al., 1992). Such techniques perform well since they do not exclusively compute geometric relationships between important points. Rather, they compare the jets or some other transform vector response around each important point. Alternate transform techniques have been based on statistical training. For example, Pentland uses the Karhunen-Loeve decomposition to generate the optimal basis for spanning mug-shot images of human faces and then uses the subsequent transform to map the faces into a lower-dimensional representation for recognition (Turk and Pentland, 1991). This technique has been also applied by (Akamatsu et al., 1992) on the Fourier-transformed images instead of the original intensity images. Recent work by (Pentland et al., 1994) involves modular eigenspaces where the optimal intensity decomposition is performed around feature points independently (eyes, nose and mouth). Moghaddam has also investigated the application of Karhunen-Loeve decomposition to statistically recognize individuals on the basis of the spectra of their dynamic Lagrangian warping into a standard template (Moghaddam et al., 1996). These transform techniques have yielded very high recognition rates and have quickly gained popularity. However, these non-feature-based techniques do not fare well under pose changes and have difficulty with natural, un-contrived face images.

In most holistic face recognition algorithms, the face needs to be either segmented or surrounded by a simple background. Furthermore, the faces presented to the algorithms need to be roughly frontal and well-illuminated for recognition to remain accurate. This is due to the algorithms' dependence on fundamentally linear or quasi-linear analysis techniques (Fourier, wavelet, Karhunen-Loeve decompositions, etc. are linear transformations). Thus, performance degrades rapidly under three-dimensional orientation changes, nonlinear illumination variation and background clutter (i.e. large, non-linear effects).

Some psychological research has proposed that certain parts of an image attract our attention more than others (Lowe, 1985). Through visual-motor experiments, research has demonstrated that fixation time and attentional resources are generally allocated to portions in a given scene, which are visually interesting. This "degree of perceptual significance" of regions in a given image allows a human observer to almost automatically discriminate between insignificant regions in a scene and interesting ones, which warrant further investigation. The ability to rapidly evaluate the level of interest in parts of a scene could benefit a face recognition system in a similar way: by reducing its search space for possible human faces. Instead of exhaustively examining each region in an image for a face-like structure, the system would only focus computational resources upon perceptually significant objects. It has been shown that three of the important factors in evaluating perceptual significance are contrast, symmetry and scale (Kelly, 1995).

Neurophysiological experiments demonstrate that the retina performs filtering which identifies contrast spatially and temporally. For instance, center surround cells at the retinal processing stage are triggered by local spatial changes in intensity (contrast) (Rodieck and Stone, 1965). In psychological tests, humans detect high contrast objects more readily than, say, objects with a similar colour to their background. A significant change in spatial intensity is referred to as an edge, boundary or a contour. Further research has shown that response to spatial contrast also varies with time (Lettvin et al., 1959). A moving edge, for example, triggers a strong response at the retinal level. Thus, contrast or intensity changes over space and time are important in vision and this has lead to the development of edge detectors and motion detectors.

Another property in estimating perceptual importance is symmetry (Reisfeld, 1994). The precise definition of symmetry in the context of attentional mechanisms is different from the intuitive concept of symmetry. Symmetry, here, represents the symmetric enclosure or the approximate encirclement of a region by contours. The appropriate arrangements of edges, which face each other to surround a region, attract the human eye to that region. Furthermore, the concept of enclosure is different from the mathematical sense of perfect closure since humans will still perceive a sense of enclosure despite gaps in boundaries that surround the region (Kelly, 1995).

Scale is also a feature, which determines a region's relative importance in a scene (Barr, 1986). It is progressively easier to detect a foreground object if it occupies a greater and greater area in our field of view. Generally, as an object is enlarged, it increases in relative importance.

For a computer based face recognition system to succeed, it must detect faces and the features that compose them despite variations each face has in the current scene (Nastar and Pentland, 1995). Faces can be anywhere in an image, at a variety of sizes, at a variety of poses and at a variety of illuminations. Although humans quickly detect the presence and location of faces and facial features from a photograph, automatic machine detection of such objects involves a complex set of operations and tests. It is uncertain exactly how humans detect faces in an image, however we can attempt to imitate the perceptual mechanisms humans seem to employ.

9.3 Intelligent Clustering of the Human Face

In our approach for face recognition, we divide a face into three different regions, which are the eyes, nose and mouth. For this reason, we need a method to automatically dive the face into the three regions. Of course, one easy solution would be to have a fixed partition of the face, but this is not a good choice due to the diversity that exists in real human faces. For this reason, we used intelligent clustering techniques to automatically divide a face according to its specific form. In this way, a specific face may have a bigger nose region and another may have a bigger mouth region, depending on each case. We show in Fig. 9.1 an illustration of the clustering approach for dividing a human face in the three above-mentioned regions.

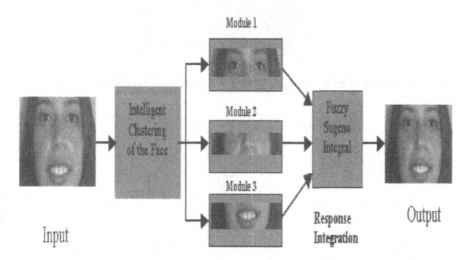

Fig. 9.1. Intelligent Clustering of the Face

As mentioned in previous chapters, there are several possible methods that can be used for performing the clustering of the face. With respect to neural networks, we have competitive networks, LVQ neural networks, and Kohonen maps. On the other hand, with respect to fuzzy logic, we have fuzzy c-means and mountain clustering. We will consider as an illustration competitive neural networks, and Kohonen maps.

9.3.1 Competitive Neural Networks for Face Clustering

As mentioned in Chap. 5, competitive networks perform clustering of the input data based on the similarities between points. We will experiments with three images with 90 × 148 pixel resolution. To perform the experiments, each image was represented as a matrix. The first image is of a student named "Paty" (shown in Fig. 9.2).

Now we show the matrix representation of the face as Fig. 9.3. In this figure, we can appreciate the distribution of points for this particular face.

Now we perform the clustering using competitive neural networks. We first show the results for 5 clusters. We show in Fig. 9.4 the results for two different cases, changing the number of epochs in the learning stage.

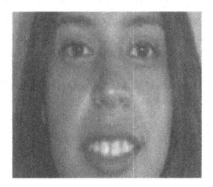

Fig. 9.2. First image used in the clustering experiment

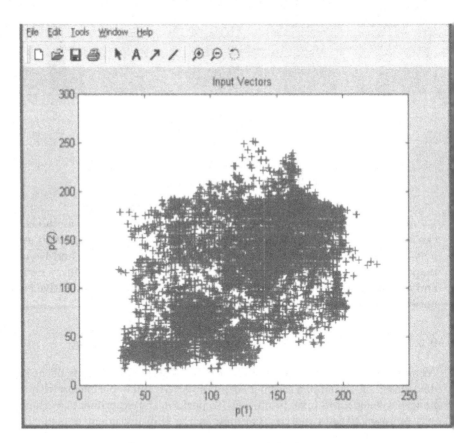

Fig. 9.3. Matrix representation of the first image

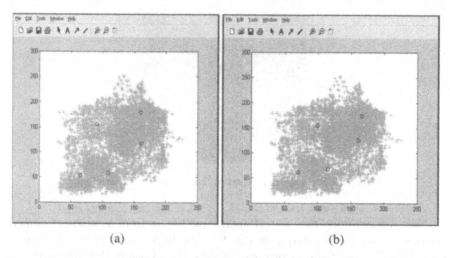

Fig. 9.4. Clustering with competitive networks for 5 clusters: **(a)** 3000 epochs, **(b)** 5000 epochs

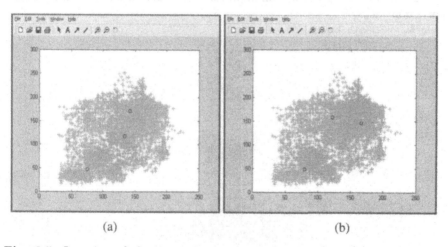

Fig. 9.5. Learning of the competitive network for first face: **(a)** 1000 epochs, **(b)** 1500 epochs

Now we consider clustering with competitive networks, but for three clusters, which is the case of interest for us because we want to divide the face into three parts. We show in Fig. 9.5 the results for two different cases, changing the number of epochs in the learning stage.

From Fig. 9.5 we can appreciate the distribution of the clusters, which will the basis for dividing this face in the three regions. Of course, due to the nature of this face the regions are not of the same size.

Fig. 9.6. Second image used in the clustering experiment

Now we consider the second image, which is from a student named "Barker". This face is shown in Fig. 9.6. We have to say that this face is very different from the previous one, as we will find out in the clustering.

Now we show the matrix representation of this second face as illustrated in Fig. 9.7. In this figure, we can appreciate the distribution of points for this particular face.

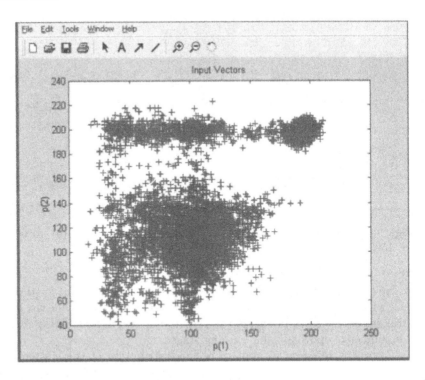

Fig. 9.7. Matrix representation of the second image

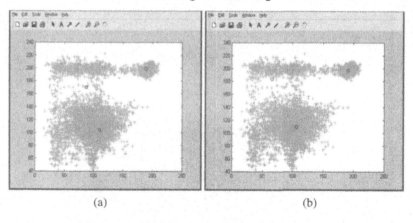

(a)   (b)

Fig. 9.8. Learning of the competitive network for second face: (**a**) 2000 epochs, (**b**) 5000 epochs

Now we consider clustering with competitive neural networks, for three clusters, which is the case of interest for us because we want to divide the face into three parts. We show in Fig. 9.8 the results for two different cases, changing the number of epochs in the learning stage.

Now we show results for the third image. This face is from a student named "Daniela". This face will again have different structure, and for this reason the distribution of clusters will different. As a consequence the division of the face into the three parts will be different with respect to the previous cases. We show in Fig. 9.9 the third image in the experiments.

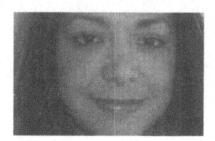

Fig. 9.9. Third image used in the clustering experiment

Now we show the matrix representation of the third face as illustrated in Fig. 9.10. In this figure, we can appreciate the distribution of points for this particular face.

Now we consider clustering with competitive neural networks, for three clusters, which is the case of interest for us because we want to divide the face into three parts. We show in Fig. 9.11 the results for two different cases, changing the number of epochs in the learning stage.

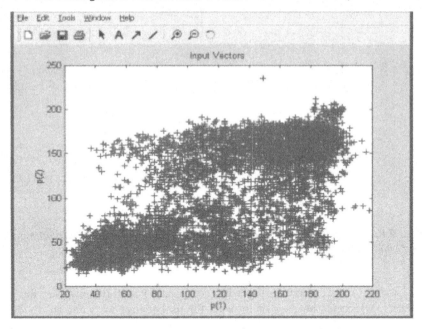

Fig. 9.10. Matrix representation of the third image

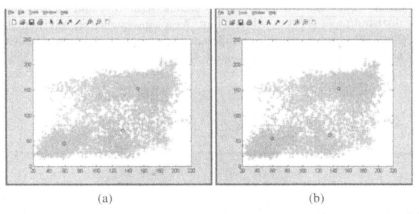

(a)   (b)

Fig. 9.11. Learning of the competitive network for third face: (a) 4000 epochs, (b) 5000 epochs

9.3.2 Kohonen Maps for Face Clustering

The goal of Kohonen maps, as seen in Chap. 5, is to organize data in classes of similar characteristics. We apply this method to the same three images as previously done with competitive networks. In this case, the results are shown as maps of points, in which the points indicate the centers of clusters that are formed.

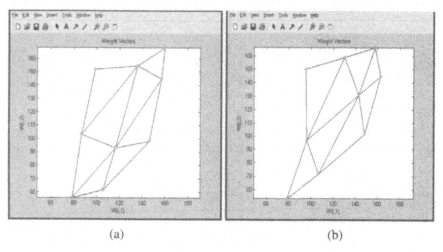

(a)   (b)

Fig. 9.12. Application of a Kohonen map for the first image: (**a**) 60 epochs, (**b**) 100 epochs

We begin by showing the results for the first image. We consider a 3 × 3 grid for finding out the structure of the face with a Kohonen map. The results are shown in Fig. 9.12.

Now we show the results for the second image. In Fig. 9.13 we illustrate the application of a Kohonen map for the second image, with different number of epochs in the learning stage.

We can appreciate from Figs. 9.13 and 9.12, that the structure in the first and second images are different. This is evident from the distribution of points in both figures.

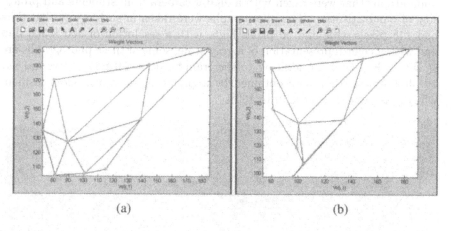

(a)   (b)

Fig. 9.13. Application of a Kohonen map for the second image: (**a**) 10 epochs, (**b**) 20 epochs

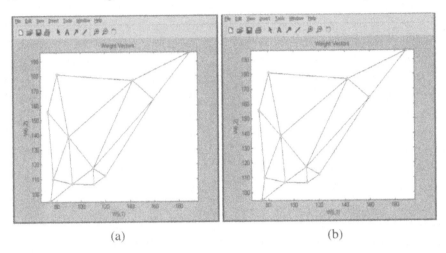

(a)   (b)

Fig. 9.14. Application of a Kohonen map for the third image: (**a**) 5 epochs, (**b**) 30 epochs

Finally we show the results for the third image. In Fig. 9.14 we illustrate the application of a Kohonen map for the third image, with different number of epochs in the learning stage.

9.4 Proposed Architecture for Face Recognition and Experimental Results

In the experiments performed in this research work, we used a total of 30 photographs that were taken with a digital camera from students and professors of our Institution. The photographs were taken in such a way that they had 148 pixels wide and 90 pixels high, with a resolution of 300 × 300 ppi, and with a color representation of a gray scale, some of these photographs are shown in Fig. 9.15. In addition to the training data (30 photos) we did use 10 photographs that were obtained by applying noise in a random fashion, which was increased from 10 to 100%.

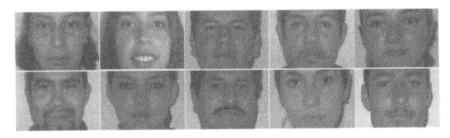

Fig. 9.15. Sample images used for training

9.4.1 Proposed Architecture for Face Recognition

The architecture proposed in this work consist of three main modules, in which each of them in turn consists of a set of neural networks trained with the same data, which provides the hierarchical modular architecture shown in Fig. 9.16.

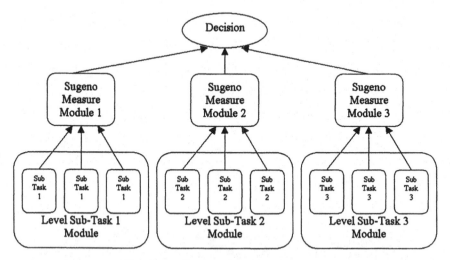

Fig. 9.16. Final proposed architecture

The input to the modular system is a complete photograph. For performing the neural network training, the images of the human faces are divided automatically by clustering into three different regions. The first region consists of the area around the eyes, which corresponds to Sub Task 1. The second region consists of the area around the nose, which corresponds to Sub Task 2. The third region consists of the area around the mouth, which corresponds to Sub Task 3. An example of this image division is shown in Fig. 9.17.

As output to the system we have an image that corresponds to the complete image that was originally given as input to the modular system, we show in Fig. 9.18 an example of this.

Fig. 9.17. Example of image division

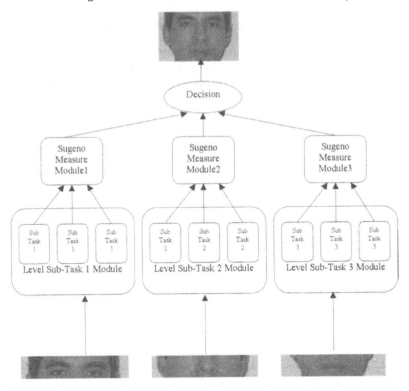

Fig. 9.18. Final architecture showing inputs and outputs

9.4.2 Description of the Integration Module

The integration modules performs its task in two phases. In the first phase, it obtains two matrices. The first matrix, called h, of dimension 3×3, stores the larger index values resulting from the competition for each of the members of the modules. The second matrix, called I, also of dimension 3×3, stores the photograph number corresponding to the particular index.

Once the first phase is finished, the second phase is initiated, in which the decision is obtained. Before making a decision, if there is consensus in the three modules, we can proceed to give the final decision, if there isn't consensus then we have search in matrix g to find the larger index values and then calculate the Sugeno fuzzy measures for each of the modules, using the following formula,

$$g(M_i) = h(A) + h(B) + \lambda h(A)h(B) \tag{9.1}$$

where λ is equal to 1. Once we have these measures, we select the largest one to show the corresponding photograph.

9.5 Summary of Results

We describe in this section the experimental results obtained with the proposed approach using the 20 photographs as training data. We show in Table 9.1 the relation between accuracy (measured as the percentage of correct results) and the percentage of noise in the figures. We also show in Fig. 9.19 a plot of the accuracy against noise.

Table 9.1. Relation between the % of noise and the % of correct results

% of Noise	% Accuracy
0	100
10	100
20	100
30	100
40	95
50	100
60	100
70	95
80	100
90	75
100	80

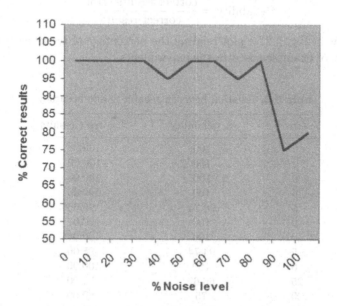

Fig. 9.19. Relation between the % of correct results and the % the applied noise

In Table 9.1 we show the relation that exists between the % of noise that was added in a random fashion to the testing data set, that consisted of the 30 original photographs, plus 300 additional images. We show in Fig. 9.20 sample images with noise.

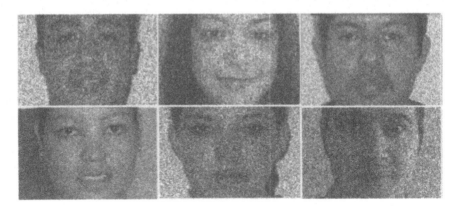

Fig. 9.20. Sample image with noise

In Table 9.2 we show the reliability results for the system. Reliability was calculated as shown in the following equation. We also show in Table 9.3 the relation between the number of examples and percentage of recognition.

$$\text{Reliability} = \frac{\text{correct results error}}{\text{correct results}} \tag{9.2}$$

We show in Fig. 9.21 a plot relating the percentage of recognition against the number of examples used in the experiments.

Table 9.2. Relation between reliability and accuracy

% errors	% reliability	% correct results
0	100	100.00
0	100	100.00
0	100	100.00
0	100	100.00
5	94.74	95.00
0	100	100.00
0	100	100.00
5	94.74	95.00
0	100	100.00
25	66.67	75.00
20	75	80.00

Table 9.3. Relation between the number of examples and the percentage of recognition

Recognized	No. of examples	% of recognition
11	11	100
20	22	91
31	33	94
42	44	95
53	55	96
64	66	97
75	77	97
86	88	98
97	99	98
108	110	98
117	121	97
127	132	96
138	143	97
148	154	96
158	165	96
169	176	96
179	187	96
190	198	96
201	209	96
210	220	95

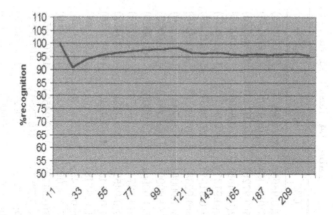

Fig. 9.21. Relation between % of recognition and number of examples

Finally, we show in Table 9.4 the results for each and every one of the training data, for each applied noise level, including the Sugeno fuzzy measure that was obtained.

Table **9.4.** Results for each of the images and different noise level applied

	Index	0%	10%	20%	30%	40%	50%	60%	70%
Barker	12	12(0.9865)	12(0.9865)	12(0.9865)	12(0.9865)	12(0.9865)	12(0.9865)	12(0.9865)	12(0.9865)
Daniela	3	3(0.9145)	3(0.9145)	3(0.9145)	3(0.9145)	3(0.9145)	3(0.9145)	3(0.9145)	3(0.9145)
Diana	5	5(0.9726)	5(0.9726)	5(0.9726)	5(0.9726)	5(0.9726)	5(0.9726)	5(0.9726)	5(0.9726)
Israel	2	2(0.8496)	2(0.8496)	2(0.8496)	2(0.8496)	2(0.8496)	2(0.8496)	2(0.8496)	2(0.8496)
Janet	13	13(0.9137)	13(0.9137)	13(0.9137)	13(0.9137)	13(0.9137)	13(0.9137)	13(0.9137)	13(0.9137)
Jocobi	14	14(0.9137)	14(0.9137)	14(0.9137)	14(0.9137)	14(0.9137)	14(0.9137)	14(0.9137)	14(0.9137)
Lapaz	6	6(0.9238)	6(0.9238)	6(0.9238)	6(0.9238)	6(0.9238)	6(0.9238)	6(0.9238)	6(0.9238)
Liliana	15	15(1.0000)	15(1.0000)	15(1.0000)	15(1.0000)	15(1.0000)	15(1.0000)	15(1.0000)	15(1.0000)
Lupita	7	7(1.0000)	7(1.0000)	7(1.0000)	7(1.0000)	7(1.0000)	7(1.0000)	7(1.0000)	7(1.0000)
Marlen	1	1(1.0000)	1(1.0000)	1(1.0000)	1(1.0000)	1(1.0000)	1(1.0000)	1(1.0000)	1(1.0000)
Papa	8	8(0.9274)	8(0.9274)	8(0.9274)	8(0.9274)	No	8(0.9274)	8(0.9274)	8(0.9274)
Paris	9	9(0.9452)	9(0.9452)	9(0.9452)	9(0.9452)	9(0.9452)	9(0.9452)	9(0.9452)	9(0.9452)
Paty	10	10(0.9146)	10(0.9146)	10(0.9146)	10(0.9146)	10(0.9146)	10(0.9146)	10(0.9146)	10(0.9146)
Pedro	11	11(0.9148)	11(0.9148)	11(0.9148)	11(0.9148)	11(0.9148)	11(0.9148)	11(0.9148)	11(0.9148)
Peludo	16	16(0.9145)	16(0.9145)	16(0.9145)	16(0.9145)	16(0.9145)	16(0.9145)	16(0.9145)	16(0.9145)
Rene	17	17(1.0000)	17(1.0000)	17(1.0000)	17(1.0000)	17(1.0000)	17(1.0000)	17(1.0000)	17(1.0000)
Sanchez	18	18(0.9452)	18(0.9452)	18(0.9452)	18(0.9452)	18(0.9452)	18(0.9452)	18(0.9452)	No
Tato	4	4(0.7881)	4(0.7881)	4(0.7881)	4(0.7881)	4(0.7881)	4(0.7881)	4(0.7881)	4(0.7881)
Toño	19	19(0.9029)	19(0.9029)	19(0.9029)	19(0.9029)	19(0.9029)	19(0.9029)	19(0.9029)	19(0.9029)
Zuñiga	20	20(1.0000)	20(1.0000)	20(1.0000)	20(1.0000)	20(1.0000)	20(1.0000)	20(1.0000)	20(1.0000)

Table 9.5. Comparison between the modular and monolithic approach

% of Noise	Modular % Correct Results	Monolithic % Correct Results
0	100	50
10	100	35
20	100	35
30	100	35
40	95	35
50	100	35
60	100	35
70	95	35
80	100	35
90	75	35
100	80	35

In addition to the experimental results presented before, we also compared the performance of the proposed modular approach, against the performance of a momolithic neural network approach. The conclusion of this comparison was that for this type of input data, the monolithic approach is not feasible, since not only training time is larger, also the recognition is too small for real-world use. We show in Table 9.5 the comparison of regonition rates between the proposed modular approach and the monolithic approach. From this table, it is easy to conclude the advantage of using the proposed modular approach for pattern recognition in human faces. We also show in Fig. 9.22 a plot showing this comparison but now in a graphical fashion.

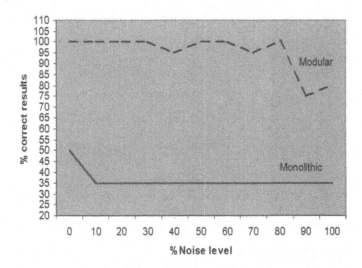

Fig. 9.22. Comparison between the modular and monolithic approach

9.6 Summary

We showed in this chapter the experimental results obtained with the proposed modular approach. In fact, we did achieve a 98.9% recognition rate on the testing data, even with an 80% level of applied noise. For the case of 100% level of applied noise, we did achieve a 96.4% recognition rate on the testing data. The testing data included 10 photographs for each image in the training data. These 10 photographs were obtained by applying noise in a random fashion, increasing the level of noise from 10 to 100%, to the training data. We also have to notice that it was achieved a 96.7% of average reliability with our modular approach. This percentage values was obtained by averaging.

In light of the results of our proposed modular approach, we have to notice that using the modular approach for human face pattern recognition is a good alternative with respect to existing methods, in particular, monolithic, gating or voting methods. As future research work, we propose the study of methods for pre-processing the data, like principal components analysis, eigenfaces, or any other method that may improve the performance of the system. Other future work includes considering different methods of fuzzy response integration, or considering evolving the number of layers and nodes of the neural network modules.

Fingerprint Recognition with Modular Neural Networks and Fuzzy Measures

We describe in this chapter a new approach for fingerprint recognition using modular neural networks with a fuzzy logic method for response integration. We describe a new architecture for modular neural networks for achieving pattern recognition in the particular case of human fingerprints. Also, the method for achieving response integration is based on the fuzzy Sugeno integral. Response integration is required to combine the outputs of all the modules in the modular network. We have applied the new approach for fingerprint recognition with a real database of fingerprints obtained from students of our institution.

10.1 Introduction

Among all the biometric techniques, fingerprint-based identification is the oldest method, which has been successfully used in numerous applications. Everyone is known to have unique, immutable fingerprints. A fingerprint is made of a series of ridges and furrows on the surface of the finger. The uniqueness of a fingerprint can be determined by the pattern of ridges and furrows as well as the minutiae points. Minutiae points are local ridge characteristics that occur at either a ridge bifurcation or a ridge ending. Fingerprint matching techniques can be placed into two categories: minutiae-based and correlation based. Minutiae-based techniques first find minutiae points and then map their relative placement on the finger. However, there are some difficulties when using this approach. It is difficult to extract the minutiae points accurately when the fingerprint is of low quality. Also this method does not take into account the global pattern of ridges and furrows. The correlation-based method is able to overcome some of the difficulties of the minutiae-based approach. However, it has some of its own shortcomings. Correlation-based techniques require the precise location of a registration point and are affected by image translation and rotation.

Patricia Melin and Oscar Castillo: *Hybrid Intelligent Systems for Pattern Recognition Using Soft Computing*, StudFuzz **172**, 207–221 (2005)
www.springerlink.com

Fingerprint matching based on minutiae has problems in matching different sized (unregistered) minutiae patterns. Local ridge structures can not be completely characterized by minutiae. We are trying an alternative representation of fingerprints, which will capture more local information and yield a fixed length code for the fingerprint. The matching will then hopefully become a relatively simple task of calculating the Euclidean distance between the two codes.

We describe algorithms, which are more robust to noise in fingerprint images and deliver increased accuracy in real-time. This is very difficult to achieve with any other technique. We are investigating methods to pool evidence from various matching techniques to increase the overall accuracy of the system. In a real application, the sensor, the acquisition system and the variation in performance of the system over time is very critical. We are also testing our system on a limited number of users to evaluate the system performance over a period of time.

The basic idea of the new approach is to divide a human fingerprint in to three different regions: the top, the middle and the bottom. Each of these regions is assigned to one module of the neural network. In this way, the modular neural network has three different modules, one for each of the regions of the human fingerprint. At the end, the final decision of fingerprint recognition is done by an integration module, which has to take into account the results of each of the modules. In our approach, the integration module uses the fuzzy Sugeno integral to combine the outputs of the three modules. The fuzzy Sugeno integral allows the integration of responses from the three modules of the top, middle and bottom of a human specific fingerprint. Other approaches in the literature use other types of integration modules, like voting methods, majority methods, and neural networks.

Response integration methods for modular neural networks that have been studied, to the moment, do not solve well real recognition problems with large sets of data or in other cases reduce the final output to the result of only one module. Also, in the particular case of fingerprint recognition, methods of weighted statistical average do not work well due to the nature of the fingerprint recognition problem. For these reasons, a new approach for fingerprint recognition using modular neural networks and fuzzy integration of responses is described in this chapter.

The new approach for fingerprint recognition was tested with a database of students and professors from our institution. This database was collected at our institution using a special scanner. The results with our new approach for fingerprint recognition on this database were excellent.

10.2 Some Basic Concepts of Fingerprint Recognition

When we interact with others we are used to identifying them by their physical appearance, their voice, or other sensory data. When we need proof of identity beyond physical appearance we obtain a signature or we look at a photo identification card. In Cyberspace, where people need to interact with digital systems or with one another remotely, we do not have these tried and true means of identification available. In almost all cases we cannot see, hear, or obtain a signature from the person with whom we are interacting. Biometrics, the measurement of a unique physical characteristic, is an ideal solution to the problem of digital identification. Biometrics makes it possible to identify ourselves to digital systems, and through these systems identify ourselves to others in Cyberspace. With biometrics we create a digital persona that makes our transactions and interactions in Cyberspace convenient and secure. Of all the biometrics available, including face, iris and retina scanning, voice identification, and others, the fingerprint is one of the most convenient and foolproof. The advantages of fingerprint biometrics for the purpose of personal digital identification include:

1. Each and every one of our ten fingerprints is unique, different from one another and from those of every other person. Even identical twins have unique fingerprints!
2. Unlike passwords, PIN codes, and smartcards that we depend upon today for identification, our fingerprints are impossible to lose or forget, and they can never be stolen.
3. We have ten fingerprints as opposed to one voice, one face or two eyes.
4. Fingerprints have been used for centuries for identification, and we have a substantial body of real world data upon which to base our claim of the uniqueness of each fingerprint. Iris scanning, for instance, is an entirely new science for which there is little or no real world data.

The skin on the inside surfaces of our hands, fingers, feet, and toes is "ridged" or covered with concentric raised patterns. These ridges are called friction ridges and they serve the useful function of making it easier to grasp and hold onto objects and surfaces without slippage. It is the many differences in the way friction ridges are patterned, broken, and forked which make ridged skin areas, including fingerprints, unique.

Fingerprints are extremely complex. In order to "read" and classify them, certain defining characteristics are used, many of which have been established by law enforcement agencies as they have created and maintained larger and larger databases of prints. We usually have two types of fingerprint characteristics for use in identification of individuals: Global Features and Local Features. Global Features are those characteristics that you can see with the naked eye. Global Features include:

- Basic Ridge Patterns
- Pattern Area
- Core Area
- Delta

- Type Lines
- Ridge Count

The Local Features are also known as Minutia Points. They are the tiny, unique characteristics of fingerprint ridges that are used for positive identification. It is possible for two or more individuals to have identical global features but still have different and unique fingerprints because they have local features – minutia points – that are different from those of others.

Large volumes of fingerprints are collected and stored everyday in a wide range of applications including forensics, access control, and driver license registration. An automatic recognition of people based on fingerprints requires that the input fingerprint be matched with a large number of fingerprints in a database. To reduce the search time and computational complexity, it is desirable to classify these fingerprints in an accurate and consistent manner so that the input fingerprint is matched only with a subset of the fingerprints in the database.

Fingerprint classification is a technique to assign a fingerprint into one of the several pre-specified types already established in the literature, which can provide an indexing mechanism. Fingerprint classification can be viewed as a coarse level matching of the fingerprints. An input fingerprint is first matched at a coarse level to one of the pre-specified types and then, at a finer level, it is compared to the subset of the database containing that type of fingerprints only.

A critical step in automatic fingerprint matching is to automatically, and reliably extract minutiae from the input fingerprint images. However, the performance of a minutiae extraction algorithm relies heavily on the quality of the input fingerprint images. In order to ensure that the performance of an automatic fingerprint identification/verification system will be robust with respect to the quality of the fingerprint images, it is essential to incorporate a fingerprint enhancement algorithm in the minutiae extraction module.

10.3 Image Pre-Processing for the Fingerprints

To improve the performance of the fingerprint recognition system, we first need to pre-process the fingerprints. The pre-processing allows to extract the most important characteristics of the fingerprint. The raw images of the fingerprints are of 200 by 198 pixels and we have a database of 50 digital fingerprins from students of our institution. For achieving this pre-processing, we used a demo version (freely available) of the VeriFinger software. This computer program allows us to open our scanned fingerprint image and extract the most important points of the fingerprint. We show in Fig. 10.1 the original fingerprint image of a particular student. We also show in Fig. 10.2 the image after processing. This processed image is more clear (has less noise) and the important points are indicated.

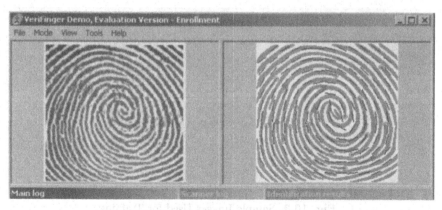

Fig. 10.1. Original image of a fingerprint

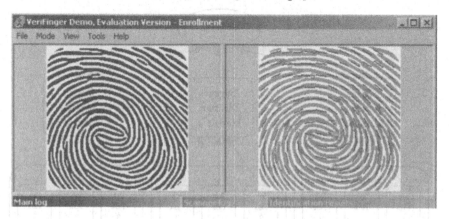

Fig. 10.2. Image of a fingerprint after processing

The processed images are the ones used as input to the neural networks. We describe in the following section how these fingerprint images are used for achieving the goal of identification.

10.4 Architecture for Fingerprint Recognition

In the experiments performed in this research work, we used 50 fingerprints that were taken with a scanner from students and professors of our Institution (Quezada, 2004). The images were taken in such a way that they had 198 pixels wide and 200 pixels high, with a resolution of 300×300 ppi, and with a color representation of a gray scale, some of these images are shown in Fig. 10.3. In addition to the training data (50 fingerprints) we did use 10 images for each fingerprint that were obtained by applying noise in a random fashion, which was increased from 10 to 100%.

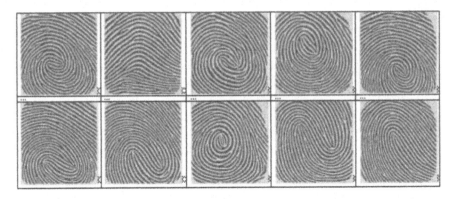

Fig. 10.3. Sample Images Used for Training

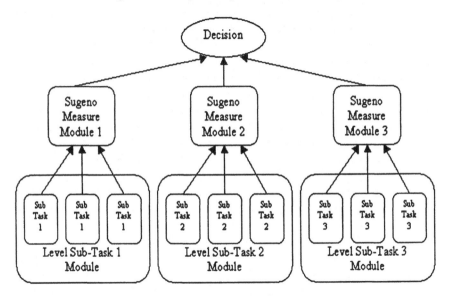

Fig. 10.4. Final Proposed Architecture

The architecture proposed for fingerprint recognition consists of three main modules, in which each of them in turn consists of a set of neural networks trained with the same data, which provides the modular architecture shown in Fig. 10.4.

The input to the modular system is a complete fingerprint image. For performing the neural network training, the images of the human fingerprints were divided in three different regions. The first region consists of the area on top, which corresponds to Sub Task 1. The second region consists of the area on the middle, which corresponds to Sub Task 2. The third region consists of the area on the bottom, which corresponds to Sub Task 3. An example of this image division is shown in Fig. 10.5.

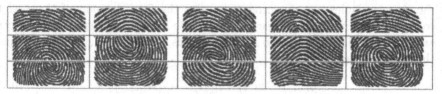

Fig. 10.5. Example of Image Division

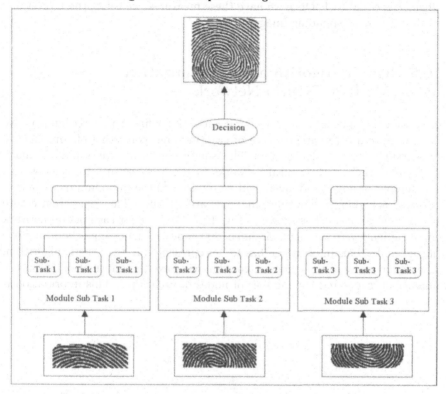

Fig. 10.6. Final architecture showing inputs and outputs

As output to the system we have an image that corresponds to the complete image that was originally given as input to the modular system, we show in Fig. 10.6 an example of this.

The integration modules performs its task in two phases. In the first phase, it obtains two matrices. The first matrix, called h, of dimension 3×3, stores the larger index values resulting from the competition for each of the members of the modules. The second matrix, called I, also of dimension 3×3, stores the image number corresponding to the particular index.

Once the first phase is finished, the second phase is initiated, in which the decision is obtained. Before making a decision, if there is consensus in the three modules, we can proceed to give the final decision, if there isn't consensus then we have to search in matrix g to find the larger index values and then calculate the Sugeno fuzzy measures for each of the modules, using the following formula,

$$g(M_i) = h(A) + h(B) + \lambda h(A)h(B) \qquad (10.1)$$

where λ is equal to 1. Once we have these measures, we select the largest one to show the corresponding image.

10.5 Genetic Algorithm for Optimization of the Modular Neural Network

To design the optimal modular neural network for fingerprint recognition, we need to optimize the architecture of each module (consisting of three neural networks) in the complete system. The genetic algorithm will simplify as much as possible each of the neural networks in the architecture. In other words, the number of nodes and layers will be reduced to the minimum necessary, in this way achieving the optimal modular architecture. The basic architecture of each neural network is shown in Fig. 10.7. We have for each neural network three hidden layers with a maximum of 200 nodes each, one input node and one output node. This information has to be represented in the genetic algorithm. The chromosome of the genetic algorithm will have the information about the number of layers and the number of nodes of each layer. This information is

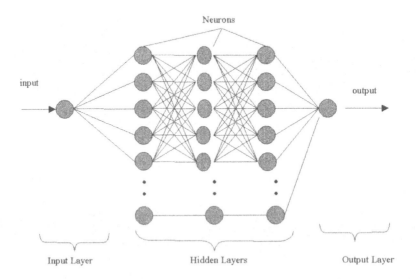

Fig. 10.7. Architecture of a neural network in the modular system

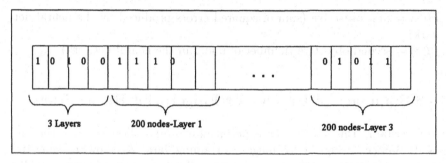

Fig. 10.8. Chromosome representation of a neural network

binary, since a 1 means the existence of a layer or a node, and a 0 means that no layer or node exists. In this way, a sequence of 0's and 1's will represent a specific architecture of a neural network. The genetic operators will be used to obtain better network architectures by the process of evolution, in which the fittest networks will survive and compete for producing new neural networks.

We show in Fig. 10.8 the chromosome representation used for an individual neural network in the genetic algorithm. We have to say that this chromosome contains hierarchical information, because the information about the layers has a higher hierarchy than the information about the nodes or neurons. This is the reason why this type of genetic algorithm is called a hierarchical genetic algorithm.

The chromosome contains the following data:

Total number of Bits = 613
Number of Layers = 3
Number of nodes in each layer = 200

The other information about the genetic algorithm is the following:

Size of the Population = 15 neural networks
Type of crossover = single-point crossover
Type of mutation: simple binary mutation
Type of selection: stochastic universal sampling

The objective function used for the genetic algorithm takes into account not only the error but also the number of nodes in the network (complexity). The basic idea is that we want good approximation, but it is also important that the size of the network is as small as possible. The following equation defines the objective function used:

$$F(z) = \alpha[\text{rank } (f_1(z))] + \beta \cdot f_2(z) . \tag{10.2}$$

Where:

$\alpha = 50$ (constant selected by the user)
$\beta = 1$ (constant selected by the user

$f1(z)$ = first objective (sum of squared errors produced by the neural network)

$f2(z)$ = second objective (number of nodes in the neural network)

10.6 Summary of Results for Fingerprint Recognition

We describe in this section the experimental results obtained with the proposed approach using the 50 images as training data. We show in Table 10.1 the relation between accuracy (measured as the percentage of correct results) and the percentage of noise in the figures.

Table 10.1. Relation between the % of noise and the % of correct results

% of Noise	% Accuracy
0	100
10	100
20	100
30	100
40	95
50	100
60	100
70	95
80	100
90	75
100	80

In Table 10.1 we show the relation that exists between the % of noise that was added in a random fashion to the testing data set, that consisted of the 50 original images, plus 500 additional images. We show in Fig. 10.9 sample images with noise.

In Table 10.2 we show the reliability results for the pattern recognition system. Reliability was calculated as shown in the following equation.

$$\text{Reliability} = \frac{\text{correct results} - \text{error}}{\text{correct results}} \tag{10.3}$$

We describe in more detail in the following lines the results of using the modular neural networks for fingerprint recognition. We first describe the results with a modular neural network that was manually designed. After that, we show the results of a modular neural network that was optimized using a hierarchical genetic algorithm.

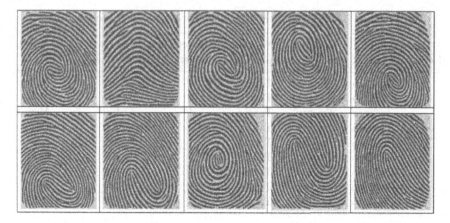

Fig. 10.9. Sample images with noise for testing

Table 10.2. Relation between reliability and accuracy

% Errors	% Reliability	% Correct Results
0	100	100.00
0	100	100.00
0	100	100.00
0	100	100.00
5	94.74	95.00
0	100	100.00
0	100	100.00
5	94.74	95.00
0	100	100.00
25	66.67	75.00
20	75	80.00

10.6.1 Recognition Results of the Modular Network without Genetic Algorithm Optimization

We first describe the results of using a modular neural network that was manually design for achieving fingerprint recognition. The number of nodes and layers in each neural network of the architecture was find through experimentation. We show in Table 10.3 the best parameters of the modular neural network, that were find through experimentation. These parameters are the result of trial and error in combination with some experience in designing neural networks for different problems. In all the experiments the number of epoch was 1000 and the error goal 0.01.

We show in Table 10.4 the detailed results for a sample of 20 fingerprints for different levels of noise. For example, fingerprint number 6 is recognized up to 60% of noise with a Sugeno measure of 0.615385. Another image that is not recognized at any level of noise is image 13.

Table 10.3. Parameters of the modular neural network manually found

Module	Type of Network	Training Function	Transfer Function	Number of Layers	Neuros in Hidden Layers	Training Error
	Backpro	Trainrp	Tansig	4	200	Mse
1	Backpro	Trainrp	Tansig	4	160	Mse
	Backpro	Trainrp	Tansig	4	160	Mse
	Backpro	Trainrp	Tansig	4	200	Mse
2	Backpro	Trainrp	Tansig	4	180	Mse
	Backpro	Trainrp	Tansig	4	180	Mse
	Backpro	Trainrp	Tansig	4	190	Mse
3	Backpro	Trainrp	Tansig	4	240	Mse
	Backpro	Trainrp	Tansig	4	190	Mse

Table 10.4. Results for each fingerprint with different levels (%) of noise applied

Finger-print	10% MS	20% MS	30% MS	40% MS	50% MS	60% MS	70% MS
1	3.043141	3.043141	3.043141	3.043141	3.043141	3.043141	3.043141
2	2.560881	2.560881	2.560881	2.560881	2.560881	2.560881	2.560881
3	2.999999	2.999999	2.999999	2.999999	2.999999	2.999999	2.999999
4	0.777779	0.777779	0.777779	0.777779	0.777779	0.777779	0.777779
5	3.000000	3.000000	3.000000	3.000000	3.000000	3.000000	3.000000
6	0.615385	0.615385	0.615385	0.615385	0.615385	0.615385	NO
7	2.999998	2.999998	NO	NO	NO	NO	NO
8	3.000002	3.000002	3.000002	3.000002	3.000002	3.000002	3.000002
9	3.000003	3.000003	3.000003	3.000003	3.000003	3.000003	3.000003
10	2.999999	2.999999	2.999999	2.999999	2.999999	2.719862	2.719862
11	NO	NO	NO	NO	NO	NO	NO
12	2.504059	2.504059	2.504059	2.504059	2.504059	2.504059	2.069608
13	NO	NO	NO	NO	NO	NO	NO
14	2.952918	2.952918	2.952918	2.952918	2.952918	2.952918	2.952918
15	2.999999	2.999999	2.999999	2.999999	2.999999	2.999999	2.999999
16	1.250000	1.250000	1.250000	1.250000	1.250000	1.250000	1.250000
17	2.384798	2.384798	2.384798	2.384798	2.384798	2.384798	2.384798
18	2.999999	2.999999	2.999999	2.999999	2.999999	2.999999	2.999999
19	2.232802	2.232802	2.232802	2.232802	2.232802	2.232802	2.232802
20	2.999999	2.999999	2.999999	2.999999	2.999999	2.999999	2.999999

We now show in Fig. 10.10 the relation between the recognition percentage and the level of noise applied. As we can appreciate from this figure the results are not as good as we expected. We have a 90% recognition rate up to a 20% of noise applied and then performance goes down to 80%. We will see in the next section that optimizing the modular network with a genetic algorithm improves these results tremendously.

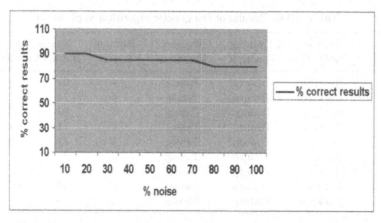

Fig. 10.10. Relation between the percentage of recognition and the level of noise applied

10.6.2 Recognition Results of the Modular Network with Genetic Algorithm Optimization

As we described previously, it is possible to use a hierarchical genetic algorithm for optimizing the architecture of the neural networks. For this reason, we applied the hierarchical genetic algorithm approach for minimizing the number of layers and nodes for all the networks in the respective modules that form the modular neural network. We show in Table 10.5 the parameters used in the genetic algorithm for optimizing the networks.

We now show in Table 10.6 the results obtained with the genetic algorithm in all of the experiments the number of generations was 1000. We can appreciate from this table that we now have the optimal number of layers and

Table 10.5. Parameters for optimizing the modular neural network

Module	Population	Number of Generations	Num. Bits	Best	Num. of Network	Num. Hidden Layers	Neurons in Hidden Layers	
	15	200	613	242	12	2	68	86
1	15	200	613	231	13	2	71	76
	15	200	613	239	3	2	72	74
	20	200	613	246	10	1	82	–
2	20	200	613	246	19	1	89	–
	20	200	613	234	14	1	79	–
	25	300	613	226	12	2	74	76
3	25	300	613	228	5	2	71	79
	25	300	613	226	12	2	74	76

Table 10.6. Results of the genetic algorithm application

Module	Type of Network	Training Function	Transfer Function	Num. of Layers	Neurons in Hidden Layers		Error Goal
	Backpro	Trainrp	Tansig	4	68	86	0.01
1	Backpro	Trainrp	Tansig	4	74	76	0.02
	Backpro	Trainrp	Tansig	3	82	–	0.01
	Backpro	Trainrp	Tansig	4	71	76	0.01
2	Backpro	Trainrp	Tansig	4	71	79	0.02
	Backpro	Trainrp	Tansig	3	89	–	0.01
	Backpro	Trainrp	Tansig	4	72	74	0.01
3	Backpro	Trainrp	Tansig	4	74	76	0.02
	Backpro	Trainrp	Tansig	3	79	–	0.01

nodes for each of the neural networks for the three modules of the complete architecture.

Now we show in Fig. 10.11 the relation between the percentage of recognition, achieved in this case, with respect to the level of noise applied. We can clearly appreciate from this figure the improvement achieved due to the application of the genetic algorithm for optimizing the modular neural network architecture. Now we have a 100% level of recognition up to 70% of noise applied, and then after this, performance goes down (see Table 10.7). It is clear that finding the optimal architecture of the networks in the modules, results in an improvement on the recognition rate of the modular neural network.

We also have to point out that due to the application of the genetic algorithm for optimizing the networks in the three modules, the training time of the neural networks was reduced considerably. Of course, this is due to the fact that the networks are smaller and as consequence require less computation time.

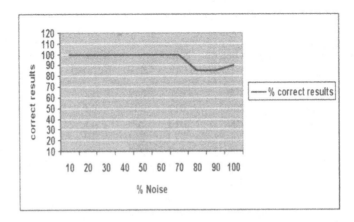

Fig. 10.11. Relation between the recognition rate and the level of noise

Table 10.7. Results for each fingerprint with different % of noise applied

Finger-print	10% MS	20% MS	30% MS	40% MS	50% MS	60% MS	70% MS
1	3.000000	3.000000	3.000000	3.000000	3.000000	3.000000	3.000000
2	2.771493	2.771493	2.771493	2.771493	2.771493	2.771493	2.461565
3	2.613652	2.613652	2.613652	2.613652	2.613652	2.613652	2.613652
4	2.907039	2.907039	2.907039	2.907039	2.907039	2.104775	2.104775
5	3.028653	3.028653	3.028653	3.028653	3.028653	3.028653	3.028653
6	3.000001	3.000001	3.000001	3.000001	3.000001	3.000001	3.000001
7	2.999997	2.999997	2.999997	2.999997	2.999997	2.999997	3.680908
8	3.000000	3.000000	3.000000	3.000000	3.000000	3.000000	3.000000
9	2.999999	2.999999	2.999999	2.999999	2.999999	2.999999	2.999999
10	2.999995	2.999995	2.999995	2.999995	2.999995	2.999995	2.999995
11	2.999998	2.999998	2.999998	2.999998	2.999998	2.999998	2.999998
12	3.000000	3.000000	3.000000	3.000000	3.000000	3.000000	3.000000
13	2.999997	2.999997	2.999997	2.999997	2.999997	2.999997	2.999997
14	3.000000	3.000000	3.000000	3.000000	3.000000	3.000000	3.000000
15	3.000001	3.000001	3.000001	3.000001	3.000001	3.000001	3.000001
16	3.000000	3.000000	3.000000	3.000000	3.000000	3.000000	3.000000
17	3.000001	3.000001	3.000001	3.000001	3.000001	3.000001	3.000001
18	3.348529	3.348529	3.348529	3.348529	3.348529	3.348529	3.348529
19	3.029244	3.029244	3.029244	3.029244	3.029244	3.029244	3.029244
20	3.000007	3.000007	3.000007	3.000007	3.000007	3.000007	3.000007

10.7 Summary

We described in this chapter the experimental results obtained with the proposed modular approach. In fact, we did achieve a 100% recognition rate on the testing data, even with an 70% level of applied noise. For the case of 100% level of applied noise, we did achieve a 90% recognition rate on the testing data. The testing data included 10 images for each fingerprint in the training data. These 10 images were obtained by applying noise in a random fashion, increasing the level of noise from 10 to 100%, to the training data. We also have to notice that it was achieved a 96.7% of average reliability with our modular approach. These percentage values were obtained by averaging. In light of the results of our proposed modular approach, we have to notice that using the modular approach for human fingerprint pattern recognition is a good alternative with respect to existing methods, in particular, monolithic, gating or voting methods. As future research work, we propose the study of methods for pre-processing the data, like principal components analysis, eigenvalues, or any other method that may improve the performance of the system. Other future work include considering different methods of fuzzy response integration, or considering evolving the number of layers and nodes of the neural network modules.

11

Voice Recognition
with Neural Networks, Fuzzy Logic
and Genetic Algorithms

We describe in this chapter the use of neural networks, fuzzy logic and genetic algorithms for voice recognition. In particular, we consider the case of speaker recognition by analyzing the sound signals with the help of intelligent techniques, such as the neural networks and fuzzy systems. We use the neural networks for analyzing the sound signal of an unknown speaker, and after this first step, a set of type-2 fuzzy rules is used for decision making. We need to use fuzzy logic due to the uncertainty of the decision process. We also use genetic algorithms to optimize the architecture of the neural networks. We illustrate our approach with a sample of sound signals from real speakers in our institution.

11.1 Introduction

Speaker recognition, which can be classified into identification and verification, is the process of automatically recognizing who is speaking on the basis of individual information included in speech waves. This technique makes it possible to use the speaker's voice to verify their identity and control access to services such as voice dialing, banking by telephone, telephone shopping, database access services, information services, voice mail, security control for confidential information areas, and remote access to computers.

Figure 11.1 shows the basic components of speaker identification and verification systems. Speaker identification is the process of determining which registered speaker provides a given utterance. Speaker verification, on the other hand, is the process of accepting or rejecting the identity claim of a speaker. Most applications in which a voice is used as the key to confirm the identity of a speaker are classified as speaker verification.

Speaker recognition methods can also be divided into text-dependent and text-independent methods. The former require the speaker to say key words or sentences having the same text for both training and recognition trials, whereas the latter do not rely on a specific text being spoken.

Patricia Melin and Oscar Castillo: *Hybrid Intelligent Systems for Pattern Recognition Using Soft Computing*, StudFuzz **172**, 223–240 (2005)
www.springerlink.com

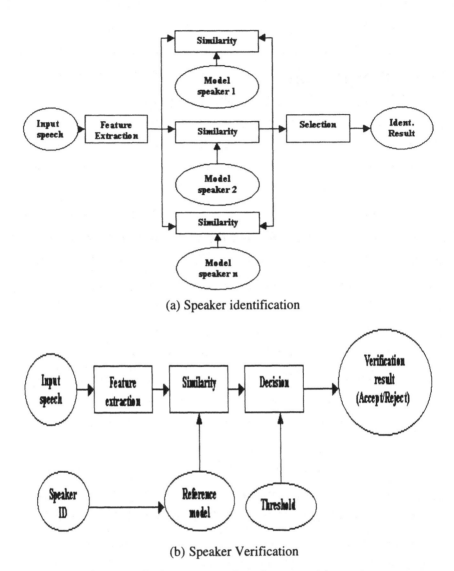

(a) Speaker identification

(b) Speaker Verification

Fig. 11.1. Basic structure of speaker recognition systems

Both text-dependent and independent methods share a problem however. These systems can be easily deceived because someone who plays back the recorded voice of a registered speaker saying the key words or sentences can be accepted as the registered speaker. To cope with this problem, there are methods in which a small set of words, such as digits, are used as key words and each user is prompted to utter a given sequence of key words that is randomly chosen every time the system is used. Yet even this method is not completely reliable, since it can be deceived with advanced electronic recording equipment that can reproduce key words in a requested order. Therefore, a text-prompted speaker recognition method has recently been proposed by (Matsui and Furui, 1993).

11.2 Traditional Methods for Speaker Recognition

Speaker identity is correlated with the physiological and behavioral character- istics of the speaker. These characteristics exist both in the spectral envelope (vocal tract characteristics) and in the supra-segmental features (voice source characteristics and dynamic features spanning several segments).

The most common short-term spectral measurements currently used are Linear Predictive Coding (LPC)-derived cepstral coefficients and their regres- sion coefficients. A spectral envelope reconstructed from a truncated set of cepstral coefficients is much smoother than one reconstructed from LPC coef- ficients. Therefore it provides a stabler representation from one repetition to another of a particular speaker's utterances. As for the regression coefficients, typically the first- and second-order coefficients are extracted at every frame period to represent the spectral dynamics. These coefficients are derivatives of the time functions of the cepstral coefficients and are respectively called the delta- and delta-delta-cepstral coefficients.

11.2.1 Normalization Techniques

The most significant factor affecting automatic speaker recognition perfor- mance is variation in the signal characteristics from trial to trial (inter-session variability and variability over time). Variations arise from the speaker them- selves, from differences in recording and transmission conditions, and from background noise. Speakers cannot repeat an utterance precisely the same way from trial to trial. It is well known that samples of the same utterance recorded in one session are much more highly correlated than samples recorded in separate sessions. There are also long-term changes in voices.

It is important for speaker recognition systems to accommodate to these variations. Two types of normalization techniques have been tried; one in the parameter domain, and the other in the distance/similarity domain.

11.2.2 Parameter-Domain Normalization

Spectral equalization, the so-called *blind equalization* method, is a typical nor- malization technique in the parameter domain that has been confirmed to be effective in reducing linear channel effects and long-term spectral variation (Furui, 1981). This method is especially effective for text-dependent speaker recognition applications that use sufficiently long utterances. Cepstral coeffi- cients are averaged over the duration of an entire utterance and the averaged values subtracted from the cepstral coefficients of each frame. Additive variation in the log spectral domain can be compensated for fairly well by this method. However, it unavoidably removes some text-dependent and speaker specific features; therefore it is inappropriate for short utterances in speaker recognition applications.

11.2.3 Distance/Similarity-Domain Normalization

A normalization method for distance (similarity, likelihood) values using a likelihood ratio has been proposed by (Higgins et al., 1991). The likelihood ratio is defined as the ratio of two conditional probabilities of the observed measurements of the utterance: the first probability is the likelihood of the acoustic data given the claimed identity of the speaker, and the second is the likelihood given that the speaker is an imposter. The likelihood ratio normalization approximates optimal scoring in the Bayes sense.

A normalization method based on a posteriori probability has also been proposed by (Matsui and Furui, 1994). The difference between the normalization method based on the likelihood ratio and the method based on a posteriori probability is whether or not the claimed speaker is included in the speaker set for normalization; the speaker set used in the method based on the likelihood ratio does not include the claimed speaker, whereas the normalization term for the method based on a posteriori probability is calculated by using all the reference speakers, including the claimed speaker.

Experimental results indicate that the two normalization methods are almost equally effective (Matsui and Furui, 1994). They both improve speaker separability and reduce the need for speaker-dependent or text-dependent thresholding, as compared with scoring using only a model of the claimed speaker.

A new method in which the normalization term is approximated by the likelihood of a single mixture model representing the parameter distribution for all the reference speakers has recently been proposed. An advantage of this method is that the computational cost of calculating the normalization term is very small, and this method has been confirmed to give much better results than either of the above-mentioned normalization methods.

11.2.4 Text-Dependent Speaker Recognition Methods

Text-dependent methods are usually based on template-matching techniques. In this approach, the input utterance is represented by a sequence of feature vectors, generally short-term spectral feature vectors. The time axes of the input utterance and each reference template or reference model of the registered speakers are aligned using a dynamic time warping (DTW) algorithm and the degree of similarity between them, accumulated from the beginning to the end of the utterance, is calculated.

The hidden Markov model (HMM) can efficiently model statistical variation in spectral features. Therefore, HMM-based methods were introduced as extensions of the DTW-based methods, and have achieved significantly better recognition accuracies (Naik et al., 1989).

11.2.5 Text-Independent Speaker Recognition Methods

One of the most successful text-independent recognition methods is based on vector quantization (VQ). In this method, VQ code-books consisting of a small number of representative feature vectors are used as an efficient means of characterizing speaker-specific features. A speaker-specific code-book is generated by clustering the training feature vectors of each speaker. In the recognition stage, an input utterance is vector-quantized using the code-book of each reference speaker and the VQ distortion accumulated over the entire input utterance is used to make the recognition decision.

Temporal variation in speech signal parameters over the long term can be represented by stochastic Markovian transitions between states. Therefore, methods using an ergodic HMM, where all possible transitions between states are allowed, have been proposed. Speech segments are classified into one of the broad phonetic categories corresponding to the HMM states. After the classification, appropriate features are selected.

In the training phase, reference templates are generated and verification thresholds are computed for each phonetic category. In the verification phase, after the phonetic categorization, a comparison with the reference template for each particular category provides a verification score for that category. The final verification score is a weighted linear combination of the scores from each category.

This method was extended to the richer class of mixture autoregressive (AR) HMMs. In these models, the states are described as a linear combination (mixture) of AR sources. It can be shown that mixture models are equivalent to a larger HMM with simple states, with additional constraints on the possible transitions between states.

It has been shown that a continuous ergodic HMM method is far superior to a discrete ergodic HMM method and that a continuous ergodic HMM method is as robust as a VQ-based method when enough training data is available. However, when little data is available, the VQ-based method is more robust than a continuous HMM method (Matsui and Furui, 1993).

A method using statistical dynamic features has recently been proposed. In this method, a multivariate auto-regression (MAR) model is applied to the time series of cepstral vectors and used to characterize speakers. It was reported that identification and verification rates were almost the same as obtained by a HMM-based method.

11.2.6 Text-Prompted Speaker Recognition Method

In the text-prompted speaker recognition method, the recognition system prompts each user with a new key sentence every time the system is used and accepts the input utterance only when it decides that it was the registered speaker who repeated the prompted sentence. The sentence can be displayed as characters or spoken by a synthesized voice. Because the vocabulary is unlimited, prospective impostors cannot know in advance what sentence will be requested. Not only can this method accurately recognize speakers, but it can also reject utterances whose text differs from the prompted text, even if it is spoken by the registered speaker. A recorded voice can thus be correctly rejected.

This method is facilitated by using speaker-specific phoneme models, as basic acoustic units. One of the major issues in applying this method is how to properly create these speaker-specific phoneme models from training utterances of a limited size. The phoneme models are represented by Gaussian-mixture continuous HMMs or tied-mixture HMMs, and they are made by adapting speaker-independent phoneme models to each speaker's voice. In order, to properly adapt the models of phonemes that are not included in the training utterances, a new adaptation method based on tied-mixture HMMs was recently proposed by (Matsui and Furui, 1994).

In the recognition stage, the system concatenates the phoneme models of each registered speaker to create a sentence HMM, according to the prompted text. Then the likelihood of the input speech matching the sentence model is calculated and used for the speaker recognition decision. If the likelihood is high enough, the speaker is accepted as the claimed speaker.

Although many recent advances and successes in speaker recognition have been achieved, there are still many problems for which good solutions remain to be found. Most of these problems arise from variability, including speaker-generated variability and variability in channel and recording conditions. It is very important to investigate feature parameters that are stable over time, insensitive to the variation of speaking manner, including the speaking rate and level, and robust against variations in voice quality due to causes such as voice disguise or colds. It is also important to develop a method to cope with the problem of distortion due to telephone sets and channels, and background and channel noises.

From the human-interface point of view, it is important to consider how the users should be prompted, and how recognition errors should be handled. Studies on ways to automatically extract the speech periods of each person separately from a dialogue involving more than two people have recently appeared as an extension of speaker recognition technology.

This section was not intended to be a comprehensive review of speaker recognition technology. Rather, it was intended to give an overview of recent advances and the problems, which must be solved in the future (Furui, 1991).

11.2.7 Speaker Verification

The speaker-specific characteristics of speech are due to differences in physiological and behavioral aspects of the speech production system in humans. The main physiological aspect of the human speech production system is the vocal tract shape. The vocal tract modifies the spectral content of an acoustic wave as it passes through it, thereby producing speech. Hence, it is common in speaker verification systems to make use of features derived only from the vocal tract.

The acoustic wave is produced when the airflow, from the lungs, is carried by the trachea through the vocal folds. This source of excitation can be characterized as phonation, whispering, frication, compression, vibration, or a combination of these. Phonated excitation occurs when the airflow is modulated by the vocal folds. Whispered excitation is produced by airflow rushing through a small triangular opening between the arytenoid cartilage at the rear of the nearly closed vocal folds. Frication excitation is produced by constrictions in the vocal tract. Compression excitation results from releasing a completely closed and pressurized vocal tract. Vibration excitation is caused by air being forced through a closure other than the vocal folds, especially at the tongue. Speech produced by phonated excitation is called voiced, that produced by phonated excitation plus frication is called mixed voiced, and that produced by other types of excitation is called unvoiced.

Using cepstral analysis as described in the previous section, an utterance may be represented as a sequence of feature vectors. Utterances spoken by the same person but at different times result in similar yet a different sequence of feature vectors. The purpose of voice modeling is to build a model that captures these variations in the extracted set of features. There are two types of models that have been used extensively in speaker verification and speech recognition systems: stochastic models and template models. The stochastic model treats the speech production process as a parametric random process and assumes that the parameters of the underlying stochastic process can be estimated in a precise, well-defined manner. The template model attempts to model the speech production process in a non-parametric manner by retaining a number of sequences of feature vectors derived from multiple utterances of the same word by the same person. Template models dominated early work in speaker verification and speech recognition because the template model is intuitively more reasonable. However, recent work in stochastic models has demonstrated that these models are more flexible and hence allow for better modeling of the speech production process. A very popular stochastic model for modeling the speech production process is the Hidden Markov Model (HMM). HMMs are extensions to the conventional Markov models, wherein the observations are a probabilistic function of the state, i.e., the model is a doubly embedded stochastic process where the underlying stochastic process is not directly observable (it is hidden). The HMM can only be viewed through another set of stochastic processes that produce the sequence of observations.

The pattern matching process involves the comparison of a given set of input feature vectors against the speaker model for the claimed identity and computing a matching score. For the Hidden Markov models discussed above, the matching score is the probability that a given set of feature vectors was generated by a specific model. We show in Fig. 11.2 a schematic diagram of a typical speaker recognition system.

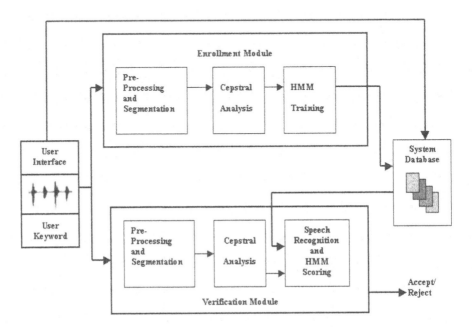

Fig. 11.2. Blocks diagram of a typical speaker recognition system

11.3 Voice Capturing and Processing

The first step for achieving voice recognition is to capture the sound signal of the voice. We use a standard microphone for capturing the voice signal. After this, we use the sound recorder of the Windows operating system to record the sounds that belong to the database for the voices of different persons. A fixed time of recording is established to have homogeneity in the signals. We show in Fig. 11.3 the sound signal recorder used in the experiments.

After capturing the sound signals, these voice signals are digitized at a frequency of 8 Khz, and as consequence we obtain a signal with 8008 sample points. This information is the one used for analyzing the voice.

Fig. 11.3. Sound recorder used in the experiments

We also used the Sound Forge 6.0 computer program for processing the sound signal. This program allows us to cancel noise in the signal, which may have come from environment noise or sensitivity of the microphones. After using this computer program, we obtain a sound signal that is as pure as possible. The program also can use fast Fourier transform for voice filtering. We show in Fig. 11.4 the use of the computer program for a particular sound signal.

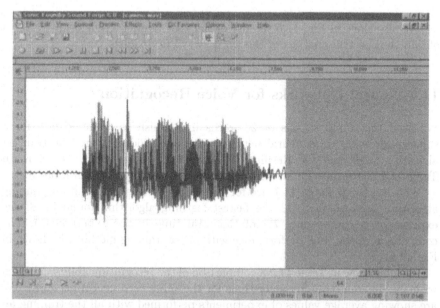

Fig. 11.4. Main window of the computer program for processing the signals

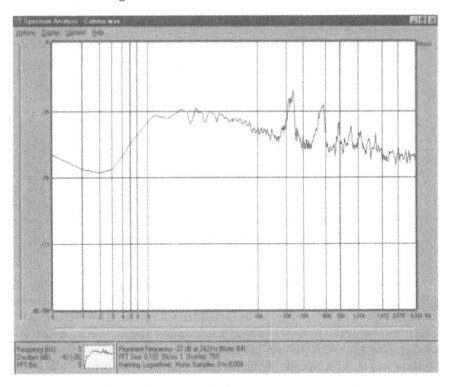

Fig. 11.5. Spectral analysis of a specific word using the FFT

We also show in Fig. 11.5 the use of the Fast Fourier Transform (FFT) to obtain the spectral analysis of the word "way" in Spanish.

11.4 Neural Networks for Voice Recognition

We used the sound signals of 20 words in Spanish as training data for a supervised feedforward neural network with one hidden layer. The training algorithm used was the Resilient Backpropagation (trainrp). We show in Table 11.1 the results for the experiments with this type of neural network.

The results of Table 11.1 are for the Resilient Backpropagation training algorithm because this was the fastest learning algorithm found in all the experiment (required only 7% of the total time in the experiments). The comparison of the time performance with other training methods is shown in Fig. 11.6.

We now show in Table 11.2 a comparison of the recognition ability achieved with the different training algorithms for the supervised neural networks. We are showing average values of experiments performed with all the training algorithms. We can appreciate from this table that the resilient backpropagation

Table 11.1. Results of feedforward neural networks for 20 words in Spanish

Stage	Time (min)	Num. of Words	No. Neurons	Words Recognized	% Recognition
1a.	11	20	50	17	85%
2a.	04	20	50	19	95%
1a.	04	20	70	16	80%
2a.	04	20	70	16	80%
3a.	02	20	25	20	100%
1a.	04	20	25	18	90%
1a.	03	20	50	18	90%
2a.	04	20	70	20	100%
2a.	04	20	50	18	90%
1a.	07	20	100	19	95%
2a.	06	20	100	20	100%
1a.	09	20	50	10	50%
1a.	07	20	75	19	95%
1a.	07	20	50	19	95%
2a.	06	20	50	20	100%
1a.	29	20	50	16	80%
1a.	43	20	100	17	85%
2a.	10	20	40	16	80%
3a.	10	20	80	16	80%
1a.	45	20	50	11	55%
2[a]	30	20	50	15	75%
3[a].	35	20	70	16	80%

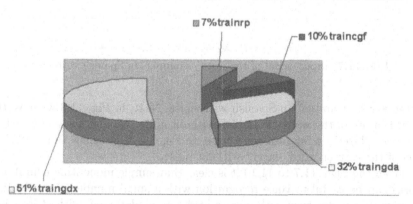

Fig. 11.6. Comparison of the time performance of several training algorithms

algorithm is also the most accurate method, with a 92% average recognition rate.

We describe below some simulation results of our approach for speaker recognition using neural networks. First, in Fig. 11.7 we have the sound signal

Table 11.2. Comparison of average recognition of four training algorithms

Method	Average Recognition
trainrp	92%
TRAINCGF-srchcha	85%
traingda	81%
traingdx	70%

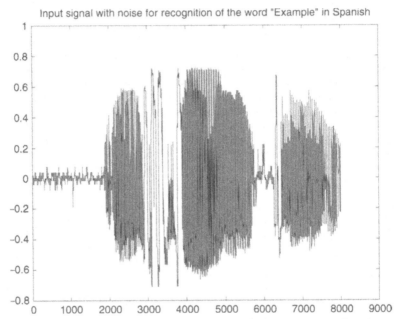

Fig. 11.7. Input signal of the word "example" in Spanish with noise

of the word "example" in Spanish with noise. Next, in Fig. 11.8 we have the identification of the word "example" without noise. We also show in Fig. 11.9 the word "layer" in Spanish with noise. In Fig. 11.10, we show the identification of the correct word "layer" without noise.

From the Figs. 11.7 to 11.10 it is clear that simple monolithic neural networks can be useful in voice recognition with a small number of words. It is obvious that words even with noise added can be identified, with at leat 92% recognition rate (for 20 words). Of course, for a larger set of words the recognition rate goes down and also computation time increases. For these reasons it is necessary to consider better methods for voice recognition.

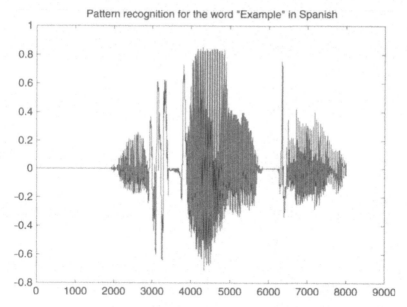

Fig. 11.8. Identification of the word "example"

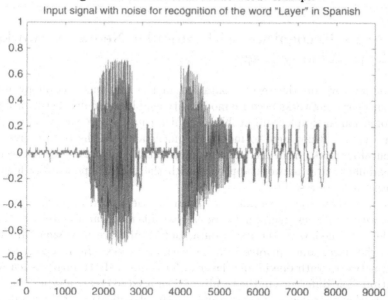

Fig. 11.9. Input signal of the word "layer" in Spanish with noise added

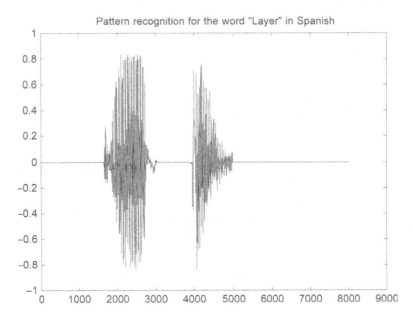

Fig. 11.10. Identification of the word "layer"

11.5 Voice Recognition with Modular Neural Networks and Type-2 Fuzzy Logic

We can improve on the results obtained in the previous section by using modular neural networks because modularity enables us to divide the problem of recognition in simpler sub-problems, which can be more easily solved. We also use type-2 fuzzy logic to model the uncertainty in the results given by the neural networks from the same training data. We describe in this section our modular neural network approach with the use of type-2 fuzzy logic in the integration of results.

We now show some examples to illustrate the hybrid approach. We use two modules with one neural network each in this modular architecture. Each module is trained with the same data, but results are somewhat different due to the uncertainty involved in the learning process. In all cases, we use neural networks with one hidden layer of 50 nodes and "trainrp" as learning algorithm. The difference in the results is then used to create a type-2 interval fuzzy set that represents the uncertainty in the classification of the word. The first example is of the word "example" in Spanish, which is shown in Fig. 11.11.

Considering for now only 10 words in the training, we have that the first neural network will give the following results:

SSE = 4.17649e-005 (Sum of squared errors)
Output = [0.0023, 0.0001, 0.0000, 0.0020, 0.0113, 0.0053, 0.0065, 0.9901, 0.0007, 0.0001]

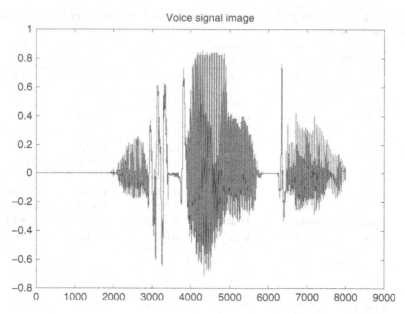

Fig. 11.11. Sound signal of the word "example" in Spanish

The output can be interpreted as giving us the membership values of the given sound signal to each of the 10 different words in the database. In this case, we can appreciate that the value of 0.9901 is the membership value to the word "example", which is very close to 1. But, if we now train a second neural network with the same architecture, due to the different random inicialization of the weights, the results will be different. We now give the results for the second neural network:

SSE = 0.0124899
Output = [0.0002, 0.0041, 0.0037, 0.0013, 0.0091, 0.0009, 0.0004, 0.9821, 0.0007, 0.0007]

We can note that now the membership value to the word "example" is of 0.9821. With the two different values of membership, we can define an interval [0.9821, 0.9901], which gives us the uncertainty in membership of the input signal belonging to the word "example" in the database. We have to use centroid deffuzification to obtain a single membership value. If we now repeat the same procedure for the whole database, we obtain the results shown in Table 11.3. In this table, we can see the results for a sample of 6 different words.

The same modular neural network approach was extended to the previous 20 words (mentioned in the previous section) and the recognition rate was improved to 100%, which shows the advantage of modularity and also the

Table 11.3. Summary of results for the two modules (M1 and M2) for a set of words in "Spanish"

Example		Daisy		Way	
M1	M2	M1	M2	M1	M2
0.0023	0.0002	0.0009	0.0124	0.0081	0.0000
0.0001	0.0041	0.9957	0.9528	0.0047	0.0240
0.0000	0.0037	0.0001	0.1141	0.0089	0.0003
0.0020	0.0013	0.0080	0.0352	0.9797	0.9397
0.0113	0.0091	0.0005	0.0014	0.0000	0.0126
0.0053	0.0009	0.0035	0.0000	0.0074	0.0002
0.0065	0.0004	0.0011	0.0001	0.0183	0.0000
0.9901	0.9821	0.0000	0.0021	0.0001	0.0069
0.0007	0.0007	0.0049	0.0012	0.0004	0.0010
0.0001	0.0007	0.0132	0.0448	0.0338	0.0007

Salina		Bed		Layer	
M1	M2	M1	M2	M1	M2
0.9894	0.9780	0.0028	0.0014	0.0009	0.0858
0.0031	0.0002	0.0104	0.0012	0.0032	0.0032
0.0019	0.0046	0.9949	0.9259	0.0000	0.0005
0.0024	0.0007	0.0221	0.0043	0.0001	0.0104
0.0001	0.0017	0.0003	0.0025	0.9820	0.9241
0.0000	0.0017	0.0003	0.0002	0.0017	0.0031
0.0006	0.0000	0.0032	0.0002	0.0070	0.0031
0.0001	0.0024	0.0003	0.0004	0.0132	0.0000
0.0067	0.0051	0.0094	0.0013	0.0003	0.0017
0.0040	0.0012	0.0051	0.0001	0.0010	0.0019

utilization of type-2 fuzzy logic. We also have to say that computation time was also reduced slightly due to the use of modularity.

We now describe the complete modular neural network architecture (Fig. 11.12) for voice recognition in which we now use three neural networks in each module. Also, each module only processes a part of the word, which is divided in three parts one for each module.

We have to say that the architecture shown in Fig. 11.12 is very similar to the ones shown in previous chapter for face and fingerprint recognition, but now the input is a voice sound signal. This signal is then divided in three parts to take advantage of the modularity, but to improve accuracy we use several simple neural networks in each module. At the end, the results of the three modules are integrated to give the final decision.

We have also experimented with using a genetic algorithm for optimizing the number of layers and nodes of the neural networks of the modules with very good results. The approach is very similar to the one described in the

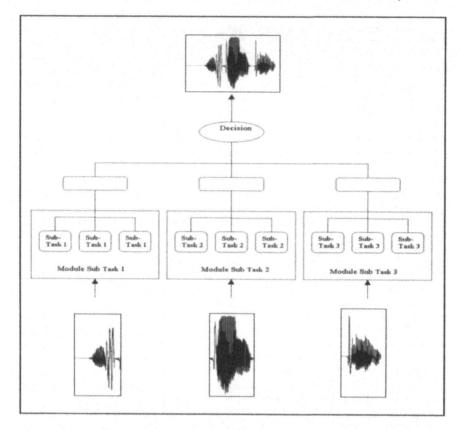

Fig. 11.12. Complete modular neural network architecture for voice recognition

previous chapter. We show in Fig. 11.13 an example of the use of a genetic algorithm for optimizing the number of layers and nodes of one of the neural networks in the modular architecture. In this figure we can appreciate the minimization of the fitness function, which takes into account two objectives: sum of squared errors and the complexity of the neural network.

11.6 Summary

We have described in this chapter an intelligent approach for pattern recognition for the case of speaker identification. We first described the use of monolithic neural networks for voice recognition. We then described a modular neural network approach with type-2 fuzzy logic. We have shown examples for words in Spanish in which a correct identification was achieved. We have performed tests with about 20 different words in Spanish, which were spoken by three different speakers. The results are very good for the monolithic

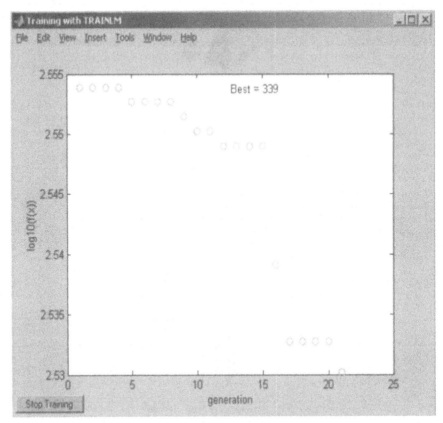

Fig. 11.13. Genetic algorithm showing the optimization of a neural network

neural network approach, and excellent for the modular neural network approach. We have considered increasing the database of words, and with the modular approach we have been able to achieve about 96% recognition rate on over 100 words. We still have to make more tests with different words and levels of noise.

12

Human Recognition using Face, Fingerprint and Voice

We describe in this chapter a new approach for human recognition using as information the face, fingerprint, and voice of a person. We have described in the previous chapters the use of intelligent techniques for achieving face recognition, fingerprint recognition, and voice identification. Now in this chapter we are considering the integration of these three biometric measures to improve the accuracy of human recognition. The new approach will integrate the information from three main modules, one for each of the three biometric measures. The new approach consists in a modular architecture that contains three basic modules: face, fingerprint, and voice. The final decision is based on the results of the three modules and uses fuzzy logic to take into account the uncertainty of the outputs of the modules.

12.1 Introduction

The term "biometrics" is derived from the Greek words bio (life) and metric (to measure). We typically choose to interpret biometrics as methods for determining unique (or relatively unique, if such an expression is allowed) features of a person's body to distinguish them from the rest of humanity. Note that there is also a branch of statistics with the same name. This branch deals with all types of data pertaining to variability in human form, some of which is of value to the "biometrics" of the computer industry.

The concept of identification based on aspects of the human body is certainly not a new one. As early as the 14th century, the Chinese were reportedly using fingerprints as a form of signature. During the late 1890s, a method of bodily measurement, called Bertillonage, (after its founder Alphonse Bertillon), was used by police authorities throughout the world. This system quickly faded when a case of two indistinguishable people with almost identical names was discovered. From this point on, fingerprinting

Patricia Melin and Oscar Castillo: *Hybrid Intelligent Systems for Pattern Recognition Using Soft Computing*, StudFuzz **172**, 241–256 (2005)
www.springerlink.com

(developed by Richard Edward Henry of Scotland Yard) became essentially the only identification tool for police.

Today, a variety of methods and techniques are available to determine unique identity, the most common being fingerprint, voice, face, and iris recognition. Of these, fingerprint and iris offer a very high level of certainty as to a person's identity, while the others are less exact. A large number of other techniques are currently being examined for suitability as identity determinants. These include (but are not limited to) retina, gait (walking style), typing style, body odour, signature, hand geometry, and DNA. Some wildly esoteric methods are also under development, such as ear structure, thermal imaging of the face and other parts of the body, subcutaneous vein patterns, blood chemistry, anti-body signatures, and heart rhythm, to name a few.

12.1.1 Major Biometric Methods

The four primary methods of biometric authentication in widespread use today are face, voice, fingerprint, and iris recognition. All of these are mentioned in this chapter, some more abundantly than others. Generally, face and voice are considered to be a lower level of security than fingerprint and iris, but on the other hand, they have a lower cost of entry. We describe briefly in this section some of these biometric methods.

Face Recognition

Facial recognition has advanced considerably in the last 10 to 15 years. Early systems, based entirely on simple geometry of key facial reference points, have given way to more advanced mathematically-based analyses such as Local Feature Analysis and Eigenface evaluation. These have been extended though the addition of "learning" systems, particularly neural networks.

The task of identifying people by face recognition can be easily divided into two sub-tasks. Firstly, there is the task of recognizing a face in a scene and extracting it from the surrounding "noise. Secondly, any computer system must then be able to extract unique features from that image and compare them with images already stored, hopefully correctly identifying the subject at hand. Based on the source of the facial image, there may also be a need to perform "liveness" tests such as determining that the head is moving against the background and that a degree of three-dimensionality can be observed. Other systems are intended to work from previously recorded images and don't test for these factors.

Face recognition systems are particularly susceptible to changes in lighting systems. For example, strong illumination from the side will present a vastly different image to a camera than neutral, evenly-positioned fluorescent lighting. Beyond this, however, these systems are relatively immune to changes such as weight gain, spectacles, beards and moustaches, and so on. Most manufacturers of face recognition systems claim false accept and false reject rates of 1% or better.

In our hybrid intelligent approach, a typical face-recognition system would rely on a simple Web camera to acquire an image prior to authentication processing, with the user gently moving his or her head while the software captures one or more images. PC resource requirements are no more than for a typical desktop PC. In an environment where moderate security is sufficient and there is no desire to purchase additional hardware (assuming the Web cameras are available), this is a very useful authentication technique.

Voice Recognition

Software systems are rapidly becoming adept at recognizing and converting free-flowing speech to its written form. The underlying difficulty in doing this is to flatten out any differences between speakers and understand everyone universally. Alternatively, when the goal is to specifically identify one person in a large group by their voice alone, these very same differences need to be identified and enhanced.

As a means of authentication, voice recognition usually takes the form of speaking a previously-enrolled phrase into a computer microphone and allowing the computer to analyze and compare the two sound samples. Methods of performing this analysis vary widely between developers. None is willing to offer more than cursory descriptions of their algorithms–principally because, apart from LAN authentication, the largest market for speaker authentication is in verification of persons over the telephone.

A number of issues contrive to weaken the performance of voice recognition as a form of biometric verification of identity. For instance, no two microphones perform identically, so the system must be flexible enough to cope with voiceprints of varying quality from a wide range of microphone performance. Also, the same person will not speak the code-phrase the same from one day to the next, or even from one minute to the next. Most researchers claim false accept and false reject rates of around 1–2%, although our research work suggests this may be excessively confident. It is safe to assume that every desktop PC in an organization is either already fitted with a microphone or can easily have one added at minimal cost. This is the main attraction of voice-based identification schemes–no other equipment or significant resources are required.

Fingerprint Recognition

The process of authenticating people based on their fingerprints can be divided into three distinct tasks. First, you must collect an image of a fingerprint; second, you must determine the key elements of the fingerprint for confirmation of identity; and third, the set of identified features must be compared with a previously-enrolled set for authentication. The system should never expect to see a complete 1:1 match between these two sets of data. In general, you could expect to couple any collection device with any algorithm, although in practice most vendors offer proprietary, linked solutions.

A number of fingerprint image collection techniques have been developed. The earliest method developed was optical: using a camera-like device to collect a high-resolution image of a fingerprint. Later developments turned to silicon-based sensors to collect an impression by a number of methods, including surface capacitance, thermal imaging, pseudo-optical on silicon, and electronic field imaging.

Unlike with the previous two methods (voice and face), it is unlikely that a typical desktop PC will be equipped with fingerprint-capture hardware. Most fingerprint reader units connect to either the parallel port (and piggy-back on the PS/2 for power) or make a USB connection. In addition to standalone units, there are fingerprint readers mounded in keyboards and in combination with smart-card readers. Recently, several laptop manufacturers have begun providing a capacitance-based fingerprint scanner either mounted next to the keyboard or as a PCMCIA-attachable unit.

The process of confirming identity via fingerprint is quick and simple for the user. Upon activation of the authentication request, most systems will energize the reader and place a viewing window on the screen. The user places a finger on the reader and can observe the quality of the captured image on the viewing window. Most systems will automatically proceed to the analysis phase upon capture of a good-quality fingerprint, but some require the user to press the <Enter> key.

The most processor-intensive portion of the recognition sequence is analyzing the scanned image to determine the location of ridges and subsequently the identification of points of termination and joining. For this phase, most manufacturers suggest a PC with a Pentium processor of at least 120 MIPS. Recent developments in smart card technology have allowed sufficient processing power to perform the actual template match on the card. This means that an enrolled template can be stored on the "hidden" side of a crypto-card and need never be released outside the card–an important step in promoting the privacy of biometric templates. However, a PC is still required to perform the image analysis.

As discussed, a variety of fingerprint detection and analysis methods exist, each with their own strengths and weaknesses. Consequently, researchers vary widely on their claimed (and achieved) false accept and false reject rates. The poorest systems offer a false accept rate of around 1:1,000, while the best are approaching 1:1,000,000. False reject rates for the same vendors are around 1:100 to 1:1000.

It is generally accepted that fingerprint recognition systems offer a moderately-priced solution with very good abilities to accurately confirm user identity. For this reason, this is the most widely-used biometric method in office environments.

Iris Recognition

Iris recognition is based entirely on a concept originated by Drs. Leonard Flom and Aran Safir, and a software process developed by Dr. John Daugman, all of Cambridge University, England. US Patent 5,291,560 issued in the name of Daugman has been assigned to Iridian Corp., one of the world's principal companies of iris-based systems. Extensive research has determined that the human iris is essentially unchanged in structure and appearance from the eighth month of gestation until a few minutes after death. Although a neonatal eye can be darkly coloured until a few months after birth, this darkness is not an influence in the infrared wavelengths normally used to collect an iris image.

Following identification and extraction of the iris from a captured image of the eye, a pseudo-polar coordinate system is established over the iris image. This allows a large number of two-dimensional modulation waveforms to be extracted that are invariant of changes due to iris widening (due to light levels) or to external factors such as spectacles or contact lenses. Iris recognition systems require special iris cameras with very high resolution and infrared illumination abilities. Typically these are attached to PCs via USB connectors. The user is required to look squarely into the camera while the eye is illuminated with a focussed infrared source. Elapsed time from capture to confirmation of identity takes less than a second. Performance requirements are no greater than those already available on the typical desktop PC.

In the history of iris recognition, there has *never* been a false acceptance. In fact, the equal error rate is 1:1,200,000, with a typical false accept rate of 1:100,000,000 and false reject rate of 1:200,000. Note that these are theoretical values based on strong analysis of limited data (only 5 to 10 million iris scans have ever been performed); they also do not take into account the perceived level of difficulty in using the system. Overall, iris scanning is the system to use if you are concerned about strongly authenticating users. The devices are considerably more expensive than fingerprint readers, but the gain in authentication confidence more than offsets the increased cost.

12.2 Biometric Systems

We describe in this section some basic concepts about biometric systems, as well as methods to evaluate their performance. We also describe the basic components of a biometric system and comparison of the different biometric measures.

Table 12.1. General comparison of biometric measures

Biometric Type	Accuracy	Ease of Use	User Acceptance
Fingerprint	High	Medium	Low
Hand Geometry	Medium	High	Medium
Voice	Medium	High	High
Retina	High	Low	Low
Iris	Medium	Medium	Medium
Signature	Medium	Medium	High
Face	Low	High	High

12.2.1 Types of Biometrics

Several different biometric modalities have emerged in recent years. The Table 12.1 lists the more common biometric sources of identity information and key characteristics of some current systems; classified in broad terms:

It is important to note that some techniques, such as retinal scanning or finger print recognition, may offer high accuracy but may not be appropriate for some applications. This is due to the high level of cooperation required by the user or the social or psychological factors that may prove unacceptable to potential users.

Both voice and face recognition are considered to be easy to use and normally acceptable by potential users. However, their accuracy is currently less than some other biometric technologies, especially in unconstrained environments such as where the background sound and illumination is variable. More information on the characteristics of specific biometric modalities can be found in (Jennings, 1992).

12.2.2 Biometric System Components and Processes

There are two distinct phases of operation for biometric systems: enrolment and verification/identification. In the first phase identity information from users is added to the system. In the second phase live biometric information from users is compared with stored records. Typical biometric identification and recognition system may have the following components:

(a) Capture: A sub-system for capturing samples of the biometric(s) to be used. This could be voice recordings or still facial images. Specific features will be extracted from the biometric samples to form templates for future comparisons. In the enrolment phase a number of such samples may be captured. A truly representative identity model may then be obtained from the features thus obtained. This enrolment process should ideally be simple and rapid, yet result in, good quality, representative templates. If the templates are of poor quality, this will affect the subsequent performance of the system. An elaborate and exhaustive enrolment process may be unacceptable.

(b) Storage: The templates thus obtained will have to be stored for future comparison. This may be done at the biometric capture device or remotely in a central server accessible via a network. Another alternative is to store the template in a portable token such as a smart card. Each one of these options has its advantages and disadvantages (Ashbourn, 1999). In addition to template storage there is often a need for a secure audit trail for all the transactions of the system.

(c) Comparison: If the biometric system is used in a verification setting, then the claimed user identity will have to be compared against the claimed reference template. The captured live biometric from the user will be compared with the claimed identity which may be provided by entering a pin, or presenting a card storing identity information. In a identification/recognition setting the live biometric will have to be compared with all the templates stored to see if there is a close match. In some systems it may be possible to automatically update the reference template after each valid match. This will make it possible for the system to adapt to gradual minor changes in user characteristics (e.g. due to aging).

(d) Interconnections: There is the need for interconnections between the capture device and the verification and storage components of the system. Often there are existing access control and information systems into which the biometric system may have to be integrated. There is a need for generic networking and programming interfaces to allow easy interconnections for biometric systems. Security and efficiency will be key considerations.

12.2.3 Biometrics System Considerations

The following are some of the key issues that need to be considered in designing and applying biometric systems.

Robustness

It is important to consider how robust the system is to fraud and impersonation. Such fraud can occur at the enrolment stage as well as at the verification stage. Using more than one biometric modality can help combat fraud and increase robustness. Also the system should be robust to small variations to the users' biometrics over time. For this, an adaptive system that gradually modifies the stored templates may be used.

Acceptability: The technology must be easy to use during both the enrolment and comparison phases. It must also be socially acceptable. The users would not accept a system that may threaten their privacy and confidentiality or that might appear to treat them as potential suspects and criminals. This accounts for the lower acceptability of fingerprint systems than voice or face recognition systems. A multimodal system is more capable to adapting to user's requirements and capabilities.

Legal issues may also have to be considered in relation to biometric systems (Woodward, 1997). There may be concerns over potential intrusions into private lives by using biometric systems. The European Union's comprehensive privacy legislation, the Directive on Data Protection, became effective on October 25, 1998. While it does not specifically mention biometrics, biometric identifiers are likely to fall within its legislative scope. The European Parliament has recently raised this issue in relation to European Community research efforts. Also, there is a growing lobby to limit and regulate the use of biometrics and surveillance technologies. Legal issues must be considered for any potential application and appropriate measures must be taken. A clear public stance on the issue of privacy in relation to biometric technologies is required to ensure broad public acceptance.

Speed and Storage Requirements

The time required to enroll, verify or identify a person is of critical importance to the acceptance and applicability of the system. Ideally, the acceptable verification time should be of the order of one second or faster. The storage requirement for the templates is also an important issue, especially if the templates are to be stored in magnetic stripe or smart cards.

Integration

The hardware platform on which the system is to be implemented is a key concern. The software, hardware and networking requirements should ideally be compatible with existing systems, allowing the biometric system to be integrated to the existing infrastructure. The system cost should be reasonable and the maintenance costs should be understood.

12.2.4 Performance Assessment

An important issue for the adoption of biometric technologies is to establish the performance of individual biometric modalities and overall systems in a credible and objective way.

For verification applications, a number of objective performance measures have been used to characterize the performance of biometric systems. In these applications a number of "clients" are enrolled onto the system. An "impostor" is defined as someone who is claiming to be someone else. The impostor may be someone who is not enrolled at all or someone who tries to claim the identity of someone else either intentionally or otherwise. When being verified the clients should be recognized as themselves and impostors should be rejected.

False Acceptance Rate (FAR) is defined as the ratio of impostors that were falsely accepted over the total number of impostors tested described as a percentage. This indicates the likelihood that an impostor may be falsely accepted and must be minimized in high-security applications.

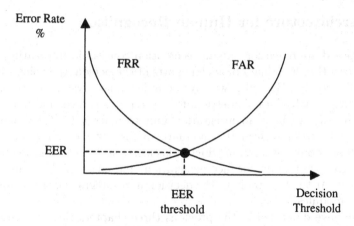

Fig. 12.1. FRR and FAR curves

False Reject Rate (FRR) is defined as the ratio of clients that are falsely rejected to the total number of clients tested described as a percentage. This indicates the probability that a valid user may be rejected by the system. Ideally this should also be minimized especially when the user community may be put-off from using the system if they are wrongly denied access.

The biometric verification process involves computing a distance between the stored template and the live sample. The decision to accept or reject is based on a pre-defined threshold. If the distance is less than this threshold then we can accept the sample. It is therefore clear that the performance of the system critically depends on the choice of this threshold and there is a trade-off between FRR and FAR. Vendors usually provide a means for controlling the threshold for their system in order to control the trade-off between FRR and FAR. The Equal Error Rate (EER) is the threshold level for which the FAR and the FRR are equal. Figure 12.1 shows a general example of the FRR and FAR curves. The EER is often quoted as a single figure to characterize the overall performance of biometric systems. Another important performance parameter is the verification time defined as the average time taken for the verification process. This may include the time taken to present the live sample.

The EU funded BIOTEST project is one initiative to provide objective performance characterization of biometric products. A National Biometric Test Centre has been established in the US and similar efforts are underway in other countries. A number of databases have been developed for the evaluation of biometric systems (Chibelushi, 1996). For the testing of joint audio-visual systems a number of research databases have been gathered in recent years (Messer et al., 1999). Developing new assessment strategies that allow meaningful comparisons between systems and solutions is an essential activity. This involves creating databases and putting together test procedures and systems for the online assessment of biometric technologies.

12.3 Architecture for Human Recognition

Our proposed approach for human recognition consists in integrating the information of the three main biometric parts of the person: the voice, the face, and the fingerprint. Basically, we have an independent system for recognizing a person from each of its biometric information (voice, face, and fingerprint), and at the end we have an integration unit to make a final decision based on the results from each of the modules. In Fig. 12.2 we show the general architecture of our approach in which it is clearly seen that we have one module for voice, one module for face recognition, and one module for fingerprint recognition. At the top, we have the decision unit integrating the results from the three modules.

As we have described in the previous three chapters the recognition systems for the face, fingerprint and voice of a human person, now we only need to concentrate in describing how to integrate the results from these three modules. The decision unit at the top of the hierarchy in Fig. 12.2 is the one that will integrate the results from the three modules. This decision unit uses fuzzy logic to take into account the uncertainty involved in the decision process.

We use, in the decision unit, a set of fuzzy rules to make final decision of human recognition. The fuzzy system has three input linguistic variables, which are face, fingerprint, and voice. Since each of these variables will have the result of the corresponding module with certain level of uncertainty, the fuzzy rules will take into account the values of the variables to give the final

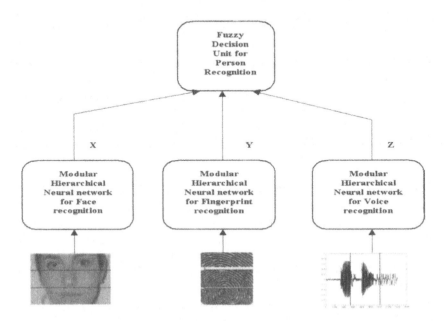

Fig. 12.2. Architecture of the proposed approach

output, which will be a specific identification of the person. We will describe the fuzzy system for decision in the following section.

12.4 Fuzzy System for the Decision on Human Recognition

We describe in this section a fuzzy system to integrate the outputs of three modules of the human recognition system. The linguistic variables of the fuzzy system are: Face, Fingerprint, and Voice. We will assume that X, Y, and Z, are three possible identifications of persons in the database. We need only to consider three possible values because we have three modules and in the worst case we will have three different results. Of course, the easiest case is when we have positive agreement in the identification of all the variables, which is the case illustrated in rule 1. The other easy case is when we have negative agreement in the three variables, which is the case illustrated in rule 9. For other cases, when we have two values in agreement and the third one is different, the output will be the majority value. When the three values are different, then the output will depend on the highest membership. Also, we can take into account that the fingerprint recognition is more reliable (according to Table 12.1), and then the voice recognition. Of course, face recognition is the least reliable of three methods used. The fuzzy rules are given as follows:

Rule 1: IF Face = X AND Fingerprint = X AND Voice = X
            THEN Person = X
Rule 2: IF Face = X AND Fingerprint = X AND Voice = Y
            THEN Person = X
Rule 3: IF Face = X AND Fingerprint = Y AND Voice = X
            THEN Person = X
Rule 4: IF Face = Y AND Fingerprint = X AND Voice = X
            THEN Person = X
Rule 5: IF Face = X AND Fingerprint = Y AND Voice = Y
            THEN Person = Y
Rule 6: IF Face = Y AND Fingerprint = Y AND Voice = X
            THEN Person = Y
Rule 7: IF Face = Y AND Fingerprint = X AND Voice = Y
            THEN Person = Y
Rule 8: IF Face = X AND Fingerprint = Y AND Voice = Z
            THEN Person = Y
Rule 9: IF Face = Z AND Fingerprint = Z AND Voice = Z
            THEN Person = Z

We now describe an implementation of this fuzzy system for verifying and validating the results of the approach for integrating the decisions of the three modules. First, we show in Fig. 12.3 the architecture of the fuzzy system,

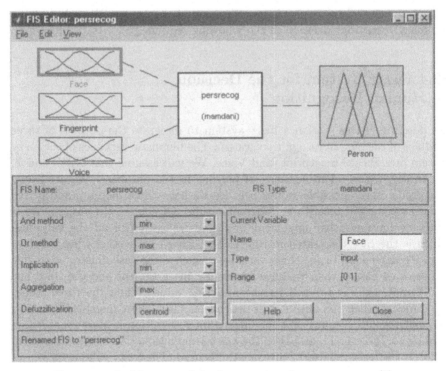

Fig. 12.3. Architecture of the fuzzy system for person recognition

which has three input linguistic variables and one output linguistic variable. We are using a Mamdani type fuzzy model with the max-min inference method and centroid deffuzzification. We also show in Figs. 12.4, 12.5, 12.6, and 12.7 the membership functions of all the linguistic variables involved in the fuzzy system. We have to note that all the membership functions are Gaussian. Regarding the membership values used in the inference, these values come from the degree of certainty of the decisions of face, fingerprint, and voice. In each case, when a decision is reached in the respective module, the final result has a degree of certainty between 0 and 1. This value is used as a membership degree in the respective linguistic value of the variable. We show in Figs. 12.8 and 12.9 two cases of identification with specific input values, which are representative of the fuzzy system use. Finally, we show in Fig. 12.10 the non-linear surface representing the fuzzy model of person recognition.

We performed extensive tests on this fuzzy system for person recognition with a database of 100 individuals from our institution (with different levels of noise, up to 100%) and the recognition rate was of about 99%, which is acceptable. Still, we can fine-tune this fuzzy system with more data, or we can improve uncertainty management by using type-2 fuzzy logic or intuitionistic fuzzy logic. We will consider these two options as future research work.

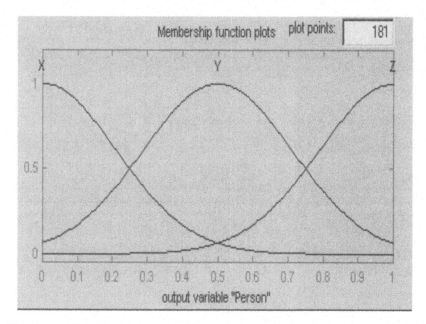

Fig. 12.4. Membership functions for the "person" linguistic output variable

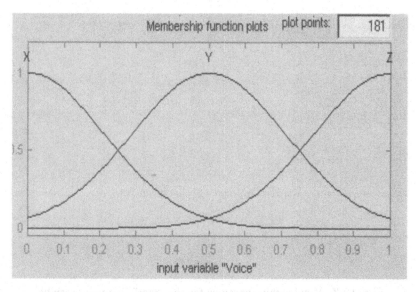

Fig. 12.5. Membership functions for the "voice" input linguistic variable

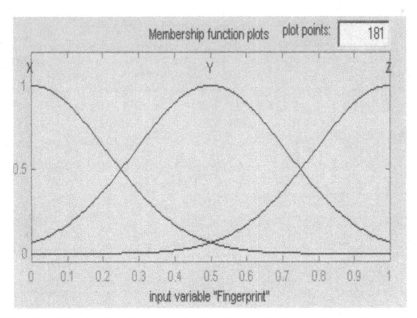

Fig. 12.6. Membership functions for the "fingerprint" input linguistic variable

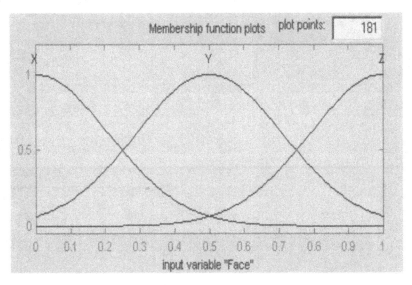

Fig. 12.7. Membership functions for the "face" linguistic variable

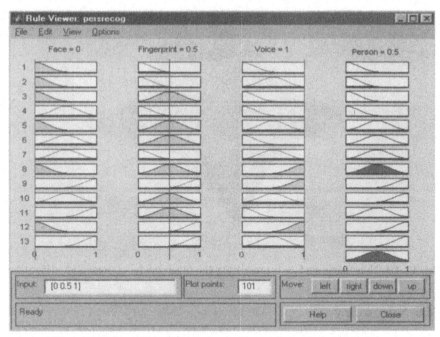

Fig. 12.8. Identification of person Y with specific input values

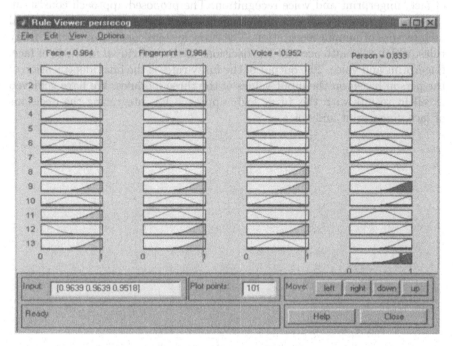

Fig. 12.9. Identification of person Z with specific input values

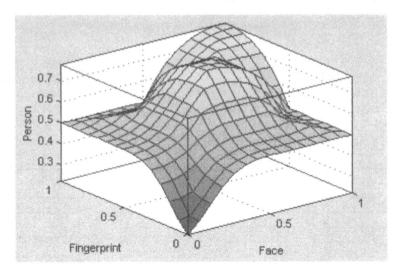

Fig. 12.10. Non-linear surface representing the fuzzy model of person identification

12.5 Summary

We described in this chapter our intelligent approach for integrating the results of face, fingerprint and voice recognition. The proposed approach consists in the use of a fuzzy system to implement the decision unit of the hierarchical architecture of human recognition. The fuzzy system consists of a set of fuzzy rules, which take into account the decisions of the individual modules of face, fingerprint and voice. The output of the fuzzy rules is the final identification of the person based on the input values of the three modules. We have achieved excellent results with this fuzzy logic approach for integrating the decisions of face, fingerprint and voice.

References

Akamatsu, S., Saski, T., Fukamachi, H., Masui, N. & Suenaga, Y. (1992). "Robust Face Identification Scheme", *Proceedings of SPIE*, pp. 71–84.

Alter, T. D. (1992). "3d pose from 3 corresponding points under weak-perspective projection", Technical Report 1378, MIT Artificial Intelligence Laboratory.

Ashbourn, J. (1999). "Practical Biometrics", Springer-Verlag, London, UK.

Atal, B. S. (1974). "Effectiveness of linear prediction characteristics of the speech wave for automatic speaker identification and verification", *Journal of the Acoustical Society of America*, Vol. 55, No. 6, pp. 1304–1312.

Atanassov, K. (1999). "Intuitionistic Fuzzy Sets: Theory and Applications", Springer-Verlag, Heidelberg, Germany.

Baker, J. E. (1987). "Reducing Bias and Inefficiency in the Selection Algorithms", Proceedings of the 2nd International Conference Genetic Algorithms, Hillsdale, NJ, pp. 14–21.

Barna, G. & Kaski, K. (1990). "Choosing optimal network structure", Proceedings of the International Neural Network Conference (INNC90), pp. 890–893.

Barr, J. M. (1986). "Speeded Phase Discrimination: Evidence for Global to Local Processing", Proceedings of Graphics Interface 1986, pp. 331–336.

Barto, A. G., Sutton, R. S. & Anderson, C. (1983). "Neuron like Elements that can Solve Difficult Learning Control Problems", *IEEE Transactions on Systems, Man & Cybernetics*, Vol. 13, pp. 835–846.

Bezdek, J.C. (1981). "Pattern Recognition with Fuzzy Objective Function Algorithms", Plenum Press, New York.

Boers, E. & Kuiper, H. (1992). "Biological Metaphors and the Design of Modular Neural Networks", Thesis, Department of Computer Science, Leiden University, The Netherlands.

Bryson, A. E. & Ho, Y.-C. (1969). "Applied Optimal Control", Blaisdell Press, New York, USA.

Cantú-Paz, E. (1995). "A Summary of Research on Parallel Genetic Algorithms", IlliGAL Report No. 95007, Illinois Genetic Algorithm Laboratory, University of Illinois at Urbana-Champaing.

Carpenter, G. A. & Grossberg, S. (1988). "The Art of Adaptive Pattern Recognition by a Self-Organizing Neural Network", *IEEE Computer*, Vol. 21 No. 3 pp. 77–88.

Castillo, O. & Melin, P. (2001). "Soft Computing for Control of Non-Linear Dynamical Systems", Springer-Verlag, Heidelberg, Germany.

Castillo, O. & Melin, P. (2002). "Hybrid Intelligent Systems for Time Series Prediction using Neural Networks, Fuzzy Logic and Fractal Theory", *IEEE Transactions on Neural Networks*, Vol. 13, No. 6, pp. 1395–1408.

Castillo, O. & Melin, P. (2003). "Soft Computing and Fractal Theory for Intelligent Manufacturing", Springer-Verlag, Heidelberg, Germany.

Castillo, O. & Melin, P. (2004). "A New Approach for Plant Monitoring using Type-2 Fuzzy Logic and Fractal Theory", *International Journal of General Systems*, Taylor and Francis, Vol. 33, pp. 305–319.

Castillo, O., Huesca, G. & Valdez, F. (2004). "Evolutionary Computing for Fuzzy System Optimization in Intelligent Control", Proceedings of IC-AI'04, Las Vegas, USA, pp. 98–104.

Chalmers, D. (1990). "The evolution of learning: an experiment in genetic connectionism". Proceedings of the 1990 Connectionist Models Summer School. Morgan Kaufmann.

Chellapa, R. (1994). "Human and Machine Recognition of Faces: A Survey", Technical Report CAR-TR-731, University of Maryland Computer Vision Laboratory.

Chen, G. & Pham, T. T. (2001). "Introduction to Fuzzy Sets, Fuzzy Logic, and Fuzzy Control Systems", CRC Press, Boca Raton, Florida, USA.

Chen, V. C. & Pao, Y. H. (1989). "Learning Control with Neural Networks", Proceedings of the International Conference on Robotics and Automation, pp. 1448–1453.

Chibelushi, C. C. (1996). "Design Issues for a Digital Audio-Visual Integrated Database", IEEE Colloqium of Integrated Audio-Visual Processing for Recognition, Synthesis and Communication, London, UK, pp. 1–7.

Cho S.-B. (2002). "Fusion of neural networks with fuzzy logic and genetic algorithms", *Journal of Integrated Computer Aided Engineering*, IOS press, Vol. 9, pp. 363–372.

Colombo, C., Del Bimbo, A., & De Magistris, S. (1995). "Human-Computer Interaction Based on Eye Movement Tracking", Computer Architectures for Machine Perception, pp. 258–263.

Cox, I. J., Ghosn, J. & Yianilos, P. N. (1995). "Feature-Based Face Recognition using Mixture-Distance", Technical Report TR-95-09, NEC Research Institute.

Craw, I., Tock, D., & Bennett, A. (1992). "Finding Face Features". Proceedings of the Second European Conference on Computer Vision, pp. 92–96.

Cybenko, G. (1989). "Approximation by Superpositions of a Sigmoidal Function", *Mathematics of Control, Signals and Systems*, Vol. 2, pp. 303–314.

DeJong, K. (1975). "The analysis and Behaviour of a Class of Genetic Adaptive Systems", PhD thesis, University of Michigan.

Denker, J. S. (1986). "Neural Networks for Computation", In J.S. Denker, editor, AIP Conference Proceedings, p. 151, American Institute of Physics.

Fahlman, S. E. (1988). "Faster-learning Variations on Back-propagation: an empirical study", Proceedings of the Conectionist Models Summer School, Carnegic Mellon University, pp. 38–51.

Fahlman, S. E. & Lebiere C. (1990). "The Cascade-Correlation Learning Architecture", Advances in Neural Information Processing Systems, Morgan Kaufmann.

Feldman, J. (1989). "Neural representation of conceptual knowledge", in Neural connections, mental computation (Nadel and et al., eds.), MIT Press.

Fogelman-Soulie, F. (1993). "Multi-modular neural network-hybrid architectures: a review", Proceedings of 1993 International Joint Conference on Neural Networks.

Fu, H.-C. & Lee, Y.-P. (2001). "Divide and Conquer Learning and Modular Perceptron Networks", *IEEE Transactions on Neural Networks*, Vol. 12, No. 2, pp. 250–263.

Fu, H.-C., Lee, Y.-P., Chiang, C.-C., & Pao, H.-T. (2001). "Divide-and-Conquer Learning and Modular Perceptron Networks", *IEEE Transactions on Neural Networks*, Vol. 12, No. 2, pp. 250–263.

Furui, S. (1981). "Cepstral analysis technique for automatic speaker verification", *IEEE Transactions on Acoustics, Speech and Signal Processing*, 29(2): pp. 254–272.

Furui, S. (1986). "Research on individuality features in speech waves and automatic speaker recognition techniques", *Speech Communication*, 5(2): pp. 183–197.

Furui, S. (1986). "Speaker-independent isolated word recognition using dynamic features of the speech spectrum", *IEEE Transactions on Acoustics, Speech and Signal Processing*, 29(1): pp. 59–59, 1986.

Furui, S. (1989). "Digital Speech Processing, Synthesis, and Recognition". Marcel Dekker, New York.

Furui, S. (1991). "Speaker-dependent-feature extraction, recognition and processing techniques", *Speech Communication*, 10(5-6): pp. 505–520.

Furui, S. (1994). "An overview of speaker recognition technology", Proceedings of the ESCA Workshop on Automatic Speaker Recognition, Identification and Verification, pp. 1–9.

Gillies, A. M. (1985). "Machine Learning Procedures for Generating Image Domain Feature Detectors", PhD thesis, University of Michigan.

Goldberg, D. E. (1989). "Genetic Algorithms in Search, Optimization and Machine Learning", Addison Wesley Publishing.

Graf, H. P., Chen, T., Petajan, E., & Cosatto, E. (1995). "Locating Faces and Facial Parts", Proceedings of the International Workshop on Automatic Face and Gesture Recognition, pp. 41–46.

Hansen, L. K. & Salamon, P. (1990). "Neural Network Ensembles", *IEEE Transactions on Pattern Analysis and Machine Intelligence*, Vol. 12, No. 10, pp. 993–1001.

Happel, B. & Murre, J. (1994). "Design and evolution of modular neural network architectures", *Neural Networks*, Vol. 7, pp. 985–1004.

Higgins, A. L., Bahler, L. & Porter, J. (1991). "Speaker verification using randomized phrase prompting", *Digital Signal Processing*, Vol. 1, pp. 89–106.

Holland, J. H. (1975). "Adaptation in Natural and Artificial Systems", University of Michigan Press.

Homaifar, A. & McCormick, E. (1995). "Simultaneous Design of Membership Functions and Rule Sets for Fuzzy Controllers using Genetic Algorithms", *IEEE Transactions on Fuzzy Systems*, Vol. 3, pp. 129–139.

Hopfield, J. J. (1982). "Neural Networks and Physical Systems with Emergent Collective Computational Abilities", Proceedings of the National Academy of Science, USA, pp. 2554–2558.

Hollstien, R. B. (1971). "Artificial Genetic Adaptation in Computer Control Systems, PhD thesis, University of Michigan.

Hsu, C. (2000). "Advanced Signal Processing Technology by Soft Computing", World Scientific Press, Singapore.

Hunt, K. J., Sbarbaro, D., Zbikowski R. & Gawthrop, P. J. (1992). "Neural Networks for Control Systems-A Survey", *Automatica*, Vol. 28 No. 6, pp. 1083–1112.

Jacobs, R. A., Jordan, M. I., Nowlan S. J. & Hinton, G. E. (1991). "Adaptive Mixtures of Local Experts", *Neural Computation*, Vol. 3, pp. 79–87.

Jamshidi, M. (1997). "Large-Scale Systems: Modelling, Control and Fuzzy Logic", Prentice-Hall.

Jain, A., Jain, A., Jain, S., & Jain, L. (2000). "Artificial Intelligence Techniques in Breast Cancer Diagnosis and Prognosis", World Scientific Press, Singapore.

Jang, J.-S. R. (1993). "ANFIS: Adaptive-Network-Based Fuzzy Inference Systems", *IEEE Transactions on Systems, Man and Cybernetics*, Vol. 23, pp. 665–685.

Jang, J.-S. R. & Gulley, N. (1997). "MATLAB: Fuzzy Logic Toolbox, User's Guide", The Math-Works, Inc. Publisher.

Jang, J.-S. R. & Sun, C.-T. (1995). "Neuro-Fuzzy Modeling and Control", *Proceedings of the IEEE*, Vol. 83, pp. 378–406.

Jang, J.-S. R., Sun, C.-T. & Mizutani, E. (1997). "Neurofuzzy and Soft Computing: A Computational Approach to Learning and Machine Intelligence", Prentice-Hall.

Janikow, C. Z. & Michalewicz, Z. (1991). "An Experimental Comparison of Binary and Floating Point Representations in Genetic Algorithms", Proceedings of 4th International Conference Genetic Algorithms, pp. 31–36.

Jenkins R. & Yuhas, B. (1993). "A simplified neural network solution through problem decomposition: The case of the truck backer-upper", *IEEE Transactions on Neural Networks*, Vol. 4, No. 4, pp. 718–722.

Jennings, N. R. (1992). "Using GRATE to Build Cooperating Agents for Industrial Control", Proceedings of IFAC/IFIPS/IMACS Symposium on Artificial Intelligence in Real Time Control, pp. 691–696.

Jordan M. & Jacobs R. (1994). "Hierarchical Mixtures of Experts and the EM Algorithm", *Journal of Neural Computations*, Vol. 5, pp. 181–214.

Kanade, T. (1973). "Picture Processing System by Computer Complex and Recognition of Human Faces", Ph.D. Thesis, Kyoto University, Japan.

Kandel, A. (1992). "Fuzzy Expert Systems", CRC Press Inc.

Kantorovich, L. V. & Akilov, G. P. (1982). "Functional Analysis", Pergamon Press, Oxford, UK.

Karnik, N. N & Mendel, J. M. (1998). "An Introduction to Type-2 Fuzzy Logic Systems", Technical Report, University of Southern California.

Karr, C. L. & Gentry, E. J. (1993). "Fuzzy Control of pH using Genetic Algorithms", *IEEE Transactions on Fuzzy Systems*, Vol. 1, pp. 46–53.

Kelly, M. F. (1995). "Annular Symmetry Operators: A Multi-Scalar for Locating and Describing Imaged Objects", Ph.D. Thesis, McGill University, Canada.

Kirkpatrick, S., Gelatt, C. D. & Vecchi, M. P. (1983). "Optimization by Simulated Annealing", *Science*, Vol. 220, pp. 671–680.

Kitano, H. (1990). "Designing neural networks using genetic algorithms with graph generation system", *Journal of Complex Systems*, Vol. 4, pp. 461–476.

Klir G. J. & Yuan, B (1995). "Fuzzy Sets and Fuzzy Logic" Theory and Applications, Prentice Hall, New York, NY, USA.

Kohonen, T. "The 'neural' phonetic typewriter", *Computer*, Vol. 21, No. 3, pp. 11–22.

Kohonen, T. (1982). "Self-Organized Formation of Topologically Correct Feature Maps, *Biological Cybernetics*, Vol. 66 pp. 59–69.

Kohonen, T. (1984). "Self-Organization and Associate Memory. Springer-Verlag, London.

Kohonen, T. (1984). "The Neural Phonetic Typewriter", *Computer*, Vol. 21 pp. 11–22.

Kohonen, T. (1989). "Self-Organization and Associate Memory. Springer-Verlag, London, 3rd Edition.

Kohonen, T. (1990). "Improved Version of Learning Vector Quantization" Proceedings of the IJCNN'90 San Diego, CA, USA, Vol. 1 pp. 545–550.

Kosko, B. (1992). "Neural Networks and Fuzzy Systems: A Dynamical Systems Approach to Machine Intelligence", Prentice-Hall.

Kosko, B. (1997). "Fuzzy Engineering", Prentice-Hall.

Kuncheva L. I. (2003). "Fuzzy" Versus "Nonfuzzy" in Combining Classifiers Designed by Boosting. *IEEE Transactions on Fuzzy Systems*, Vol. 11, No. 6, pp. 729–741.

Lee, S. & Kil, R. M. (1991). "A Gaussian Potential Function Network with Hierarchically Self-Organizing Learning", *Neural Networks* Vol. 4 No. 2, pp. 207–224.

Lettvin, J. Y., Maturana, H. R., McCulloch, W. S., and Pitts, W. H. (1959). "What the Frog's Eye tells the Fog's Brain", *Proceedings of IRE*, Vol. 47, pp. 1940–1959.

Lippmann, R. P. (1987). "An Introduction to Computing with Neural Networks", *IEEE Acoustics, Speech, and Signal Processing Magazine*, Vol. 4, pp. 4–22.

Lowe, D. (1989). "Adaptive Radial Basis Function Nonlinearities, and the Problem of Generalization, Proceedings of the First IEEE International Conference on Artificial Neural Networks, London, pp. 171–175.

Lowe, D. G. (1985). "Perceptual Organization and Visual Recognition", Kluwer Academic Publishers.

Lozano, A. (2004). "Optimizacion de un Sistema de Control Difuso por medio de Algoritmos Geneticos Jerarquicos", Thesis, Dept. of Computer Science, Tijuana Institute of Technology, Mexico.

Lu, B.-L. & Ito, M. (1998). "Task Decomposition and Module Combination Based on Class Relations: Modular Neural Network for Pattern Classification", Technical Report, Nagoya, Japan.

Lu, B.-L. & Ito, M. (1999). "Task Decomposition and Module Combination Based on Class Relations: A Modular Neural Network for Pattern Classification", *IEEE Transactions on Neural Networks*, Vol. 10, No. 5, pp. 1244–1256.

Mamdani, E. H. (1974). "Application of fuzzy algorithms for control of a simple dynamic plant", *Proceedings of the IEEE*, Vol. 121, pp. 1585–1588.

Mamdani, E. H. & Assilian, S. (1975). "An Experiment in Linguistic Synthesis with a Fuzzy Logic Controller", *International Journal of Man-Machine Studies*, Vol. 7, pp. 1–13.

Man, K. F., Tang, K. S. & Kwong, S. (1999). "Genetic Algorithms". Springer-Verlag.

Manjunath, B. S., Chellappa, R., & Malsburg, C. V. D. (1992). "A Feature Based Approach to Face Recognition" Proceedings of the IEEE Computer Society Conference on Computer Vision, pp. 373–378.

Margaliot, M. & Langholz, G. (2000). "New Approaches to Fuzzy Modeling and Control", World Scientific Press, Singapore.

Masters, T. (1993). "Practical Neural Network recipe in C++", Academic Press, Inc.

Matsui, T. & Furui, S. (1993). "Comparison of text-independent speaker recognition methods using VQ-distortion and discrete/continuous HMMs", Proceedings of ICASSP'93, pp. 157–160.

Matsui, T. & Furui, S. (1993). "Concatenated phoneme models for text-variable speaker recognition", Proceedings of ICASSP'93, pp. 391–394.

Matsui, T. and Furui, S. (1994). "Similarity normalization method for speaker verification based on a posteriori probability", Proceedings of the ESCA Workshop on Automatic Speaker Recognition, Identification and Verification, pp. 59–62.

Matsui, T. and Furui, S. (1994). "Speaker adaptation of tied-mixture-based phoneme models for text-prompted speaker recognition. Proceedings of ICASSP'94, pp. 125–128.

McCullagh, P. & Nelder, J. A. (1994). "Generalized Linear Models", Chapman and Hall, New York, USA.

Melin, P. & Castillo, O. (2002) "Modelling Simulation and Control of Non-Linear Dynamical Systems: An Intelligent Approach Using Soft Computing and Fractal Theory", Taylor and Francis Publishers, London, United Kingdom.

Melin, P., Acosta, M. L. & Felix, C. (2003). "Pattern Recognition Using Fuzzy Logic and Neural Networks", Proceedings of IC-AI'03, Las Vegas, USA, pp. 221–227.

Melin, P. & Castillo, O. (2004). "A New Method for Adaptive Control of Non-Linear Plants Using Type-2 Fuzzy Logic and Neural Networks", *International Journal of General Systems*, Taylor and Francis, Vol. 33, pp. 289–304.

Melin, P., Gonzalez, F. & Martinez, G. (2004). "Pattern Recognition Using Modular Neural Networks and Genetic Algorithms", Proceedings of IC-AI'04, Las Vegas, USA, pp. 77–83.

Melin, P., Mancilla, A., Gonzalez, C. & Bravo, D. (2004). "Modular Neural Networks with Fuzzy Sugeno Integral Response for Face and Fingerprint Recognition", Proceedings of IC-AI'04, Las Vegas, USA, pp. 91–97.

Mendel, J. (2001). "Uncertain Rule-Based Fuzzy Logic Systems: Introduction and New Directions", Prentice-Hall, New Jersey, USA.

Messer, K., Matas, J., Kittler, J., Luettin, J. and Maitre, G. (1999). "XM2VTS: The Extended M2VTS Database", Proceedings of the International Conference on Audio and Video-based Biometric Person Authentication AVBPA'99", Springer-Verlag, New York, USA, pp. 72–77.

Michalewicz, Z. (1996). "Genetic Algorithms + Data Structures = Evolution program", 3rd Edition, Springer-Verlag.

Miller, G., Todd, P. & Hedge, S. (1989). "Designing Neural Networks using Genetic Algorithms", Proceedings of the Third International Conference on Genetic Algorithms", Morgan Kauffmann.

Miller, W. T., Sutton, R. S. & Werbos P. J. (1995). "Neural Networks for Control", MIT Press.

Minsky, M. & Papert, S. (1969). "Preceptrons", MIT Press.

Mitchell, M. (1996). "An Introduction to Genetic Algorithms", MIT Press, Cambridge, MA, USA.

Mitchell, M. (1998). "An Introduction to Genetic Algorithms", MIT Press, Cambridge, USA.

Moghaddam, B., Nastar, C., & Pentland A. (1996). "Bayesian Face Recognition Using Deformable Intensity Surfaces", Technical Report 371, MIT Media Laboratory, Perceptual Computing Section.

Monrocq, C. (1993). "A probabilistic approach which provides and adaptive neural network architecture for discrimination", Proceedings of the International Conference on Artificial Neural Networks, Vol. 372, pp. 252–256.

Montana, G. & Davis, L. (1989). "Training feedforward networks using genetic algorithms", Proceedings of the International Joint Conference on Artificial Intelligence, Morgan Kauffmann.

Moody, J. & Darken, C. (1989). "Fast Learning in Networks of Locally-Tuned Processing Units", Neural Computation, Vol. 1, pp. 281–294.

Moulet, M. (1992). "A Symbolic Algorithm for Computing Coefficients Accuracy in Regression", Proceedings of the International Workshop on Machine Learning, pp. 332–337.

Mühlenbein, H. & Schilierkamp, V. (1993). "Predictive Model for Breeder Genetic Algorithm". Evolutionary Computation. 1(1): 25–49.

Murray-Smith, R. & Johansen, T. A. (1997). "Multiple Model Approaches for Modeling and Control", Taylor and Francis, United Kingdom.

Naik, J. M., Netsch, L. P. & Doddington, G. R. (1989). "Speaker verification over long distance telephone lines", Proceedings of ICASSP'89, pp. 524–527.

Nakamura, S. (1997). "Numerical Analysis and Graphic Visualization with MATLAB", Prentice Hall.

Nastar, C., & Pentland, A. (1995). "Matching and Recognition Using Deformable Intensity Surfaces", Proceedings of the International Symposium on Computer Vision, pp. 223–228.

Nixon, M. (1985). "Eye Spacing Measurement for Facial Recognition", SPIE Proceedings, pp. 279–285.

Parker D. B. (1982). "Learning Logic", Invention Report S81-64, File 1, Office of Technology Licensing.

Pentland, A., Moghaddam, B., & Starner, T. (1994). "View-Based and Modular Eigenspaces for Face Recognition", Proceedings of the IEEE Conference on Computer Vision and Pattern Recognition, pp. 84–91.

Pham, D. T. & Xing, L. (1995). "Neural Networks for Identification, Prediction and Control", Springer-Verlag.

Pomerleau, D. A. (1991). "Efficient Training of Artificial Neural Networks for Autonomous Navigation", Journal of Neural Computation, Vol. 3, pp. 88–97.

Powell, M. J. D. (1987). "Radial Basis Functions for Multivariable Interpolation: a review", In J. C. Mason & M. G. Cox, editors, Algorithms for approximation, pp. 143–167, Oxford University Press.

Procyk, T. J. & Mamdani, E. M. (1979). "A Linguistic Self-Organizing Process Controller", Automatica, Vol. 15, No. 1, pp. 15–30.

Psaltis, D., Sideris, A. & Yamamura, A. (1988). "A Multilayered Neural Network Controller", IEEE Control Systems Magazine, Vol. 8, pp. 17–21.

O'Shaughnessy, D. (1986). "Speaker recognition", *IEEE Acoustics, Speech and Signal Processing Magazine*, Vol. 3, No. 4, pp. 4–17.

Quezada, A. (2004). "Reconocimiento de Huellas Digitales Utilizando Redes Neuronales Modulares y Algoritmos Geneticos", Thesis of Computer Science, Tijuana Institute of Technology, Mexico.

Rao, R. & Ballard, D. (1995). "Natural Basis Functions and Topographics Memory for Face Recognition", Proceedings of the International Joint Conference on Artificial Intelligence, pp. 10–17.

Reisfeld, D. (1994). "Generalized Symmetry Transforms: Attentional Mechanisms and Face Recognition", Ph.D. Thesis, Tel-Aviv University, Israel.

Reisfeld, D. & Yeshurun, Y. (1992). "Robust Detection of Facial Features by Generalized Symmetry", Proceedings of the IAPR International Conference on Pattern Recognition, pp. 117–120.

Ritter, H. & Schulten, K. (1987). "Extending Kohonen's Self-Organizing Mapping Algorithm to Learn Ballistic Movements", In Rolf Eckmiller & Christoph v.d. Malsburg, editors, Neural Computers, pp. 393–406, Springer-Verlag, London.

Rodieck, R. W. & Stone, J. (1965). "Response of Cat Retinal Ganglion Cells to Moving Visual Patterns", *Journal of Neurophysiology*, Vol. 28, No. 5, pp. 819–832.

Roeder, N. & Li, X. (1995). "Face Contour Extraction from Front-View Images", *Pattern Recognition*, Vol. 28, No. 8, pp. 1167–1179.

Ronco, E. & Gawthrop, P. J. (1995). "Modular Neural Networks: State of the Art", Technical Report CSC-95026, Faculty of Engineering, University of Glasgow, United Kingdom.

Rosenberg, A. E. & Soong, F. K. (1991). "Recent research in automatic speaker recognition", In S. Furui and M. M. Sondhi, editors, *Advances in Speech Signal Processing*, Marcel Dekker, New York, pp. 701–737.

Rosenblatt, F. (1962). "Principles of Neurodynamics: Perceptrons and the Theory of Brain Mechanisms", Spartan.

Rumelhart, D. E., Hinton, G. E. & Williams, R. J. (1986). "Learning Internal Representations by Error Propagation", Parallel Distributed Processing: Explorations in the Microstructure of Cognition, Vol. 1, Chap. 8, pp. 318–362, MIT Press, Cambridge, MA, USA.

Rumelhart, D. E., & Zipser, D. (1986). "Feature Discovery by Competitive Learning", In D. E. Rumelhart and James L. McClelland, editors, Parallel Distributed Processing: Explorations in the Microstructure of Cognition, Vol. 1, Chap. 5, pp. 151–193, MIT Press, Cambridge, MA, USA.

Runkler, T. A. & Glesner, M. (1994). "Defuzzification and Ranking in the Context of Membership Value Semantics, Rule Modality, and Measurement Theory", Proceedings of European Congress on Fuzzy and Intelligent Technologies.

Ruspini, E. H. (1982). "Recent Development in Fuzzy Clustering", Fuzzy Set and Possibility Theory, North Holland, pp. 133–147.

Sackinger, E., Boser, B. E., Bromley, J., LeCun, Y. & Jackel, L. D. (1992). "Application of the Anna Neural Network Chip to High-Speed Character Recognition", *IEEE Transactions on Neural Networks*, Vol. 3, pp. 498–505.

Schmidt, A. (1996). "Implementation of a Multilayer Backpropagation Network", Master's Thesis, Department of Computing, Manchester Metropolitan University, UK.

Schmidt, A. & Bandar, Z. (1997). "A Modular Neural Network Architecture with Additional Generalization Capabilities for Large Input Vectors", Proceedings of the International Conference on Artificial Neural Networks and Genetic Algorithms (ICANNGA), Norwich, England.

Sejnowski, T. J. & Rosenberg, C. R. (1987). "Parallel Networks that Learn to Pronounce English Text", *Journal of Complex Systems*, Vol. 1, pp. 145–168.

Sharkey, A. (1999). "Combining Artificial Neural Nets: Ensemble and Modular Multi-Net Systems", Springer-Verlag, London, Great Britain.

Simon, H. (1981). "The Sciences of the Artificial", MIT Press, Cambridge, USA.

Soucek, B. (1991). "Neural and Intelligent Systems Integration: Fifth and Sixth Generation Integrated Reasoning Information Systems", John Wiley and Sons.

Spears, W. M. & DeJong, K. (1991). "An Analysis of Multi-point Crossover", In G. J. E. Rawlins, editor Foundations of Genetic Algorithms, pp. 301–315.

Staib, W. E. (1993). "The Intelligent Arc Furnace: Neural Networks Revolutionize Steelmaking", Proceedings of the World Congress on Neural Networks, pp. 466–469.

Staib, W. E. & Staib, R. B. (1992). "The Intelligent Arc Furnace Controller: A Neural Network Electrode Position Optimization System for the Electric Arc Furnace", Proceedings of the International Conference on Neural Networks, Vol. 3, pp. 1–9.

Su, H. T. & McAvoy, T. J. (1993). "Neural Model Predictive Models of Nonlinear Chemical Processes", Proceedings of the 8th International Symposium on Intelligent Control, pp. 358–363.

Su, H. T., McAvoy, T. J. & Werbos, P. J. (1992). "Long-term Predictions of Chemical Processes using Recurrent Neural Networks: A Parallel Training Approach", Industrial & Engineering Chemistry Research, Vol. 31, pp. 1338–1352.

Sugeno, M. (1974). "Theory of Fuzzy Integrals and its Applications", Ph.D. Thesis, Tokyo Institute of Technology, Japan.

Sugeno, M. & Kang, G. T. (1988). "Structure Identification of Fuzzy Model", *Journal of Fuzzy Sets and Systems*, Vol. 28, pp. 15–33.

Szmidt, E. & Kacprzyk, J. (2002). "Analysis of Agreement in a Group of experts via Distances between Intuitionistic Preferences", Proceedings of IPMU'2002, Universite Savoie, France, pp. 1859–1865.

Takagi, T. & Sugeno, M. (1985). "Fuzzy Identification of Systems and its Applications to Modeling and Control", *IEEE Transactions on Systems, Man and Cybernetics*, Vol. 15, pp. 116–132.

Tang, K.-S., Man, K.-F., Liu, Z.-F., & Kwong, S. (1998). "Minimal Fuzzy Memberships and Rules using Hierarchical Genetic Algorithms", *IEEE Transactions on Industrial Electronics*, Vol. 45, No. 1, pp. 162–169.

Troudet, T. (1991). "Towards Practical Design using Neural Computation", Proceedings of the International Conference on Neural Networks, Vol. 2, pp. 675–681.

Tsoukalas, L. H. & Uhrig, R. E. (1997). "Fuzzy and Neural Approaches in Engineering", John Wiley & Sons, Inc., New Jersey, USA.

Tsukamoto, Y. (1979). "An Approach to Fuzzy Reasoning Method", In Gupta, M. M., Ragade, R. K. and Yager, R. R., editors, Advanced in Fuzzy Set Theory and Applications, pp. 137–149, North-Holland.

Turk, M. A, and Pentland, A. P. (1991). "Face Recognition using Eigenfaces", Proceedings of the IEEE Conference on Computer Vision and Pattern Recognition, pp. 586–591.

Von Altrock, C. (1995). "Fuzzy Logic & Neuro Fuzzy Applications Explained", Prentice Hall.

Wagenknecht, M. & Hartmann, K. (1988). "Application of Fuzzy Sets of Type 2 to the solution of Fuzzy Equation Systems", *Fuzzy Sets and Systems*, Vol. 25, pp. 183–190.

Wang, L.-X. (1997). "A Course in Fuzzy Systems and Control", Prentice-Hall, Upper Saddle River, NJ, USA.

Werbos, P. J. (1991). "An Overview of Neural Networks for Control", *IEEE Control Systems Magazine*, Vol. 11, pp. 40–41.

Werbos, P. J. (1974). "Beyond Regression: New Tools for Prediction and Analysis in the Behavioral Sciences", Ph.D. Thesis, Harvard University.

Widrow, B. & Stearns, D. (1985). "Adaptive Signal Processing", Prentice-Hall.

White, H. (1989). "An Additional Hidden Unit Test for Neglected Nonlinearity in Multilayer Feedforward Networks", Proceedings of IJCNN'89, Washington, D.C., IEEE Press, pp. 451–455.

Woodward, J. D. (1997). "Biometrics: Privacy's Foe or Privacy's Friend?", *Proceedings of the IEEE*, Vol. 85, No. 9, pp. 1480–1492.

Wright, A. H. (1991). "Genetic Algorithms for Real Parameter Optimization", In J. E. Rawlins, editor, Foundations of Genetic Algorithms, Morgan Kaufmann, pp. 205–218.

Yager, R. R. (1980). "Fuzzy Subsets of Type II in Decisions", *Journal of Cybernetics*, Vol. 10, pp. 137–159.

Yager, R. R. (1993). "A General Approach to Criteria Aggregation Using Fuzzy Measures", *International Journal of Man-Machine Studies*, Vol. 1, pp. 187–213.

Yager R. R. (1999). Criteria Aggregations Functions Using Fuzzy Measures and the Choquet Integral, *International Journal of Fuzzy Systems*, Vol. 1, No. 2.

Yager, R. R. & Filev, D. P. (1993). "SLIDE: A Simple Adaptive Defuzzification Method", *IEEE Transactions on Fuzzy Systems*, Vol. 1, pp. 69–78.

Yager, R. R. & Filev, D. P. (1994). "Generation of Fuzzy Rules by Mountain Clustering", *Journal of Intelligent and Fuzzy Systems*, Vol. 2, No. 3, pp. 209–219.

Yen, J., Langari, R. & Zadeh, L. A. (1995). "Industrial Applications of Fuzzy Control and Intelligent Systems", IEEE Computer Society Press.

Yen, J. & Langari, R. (1999). "Fuzzy Logic: Intelligence, Control, and Information", Prentice Hall, New Jersey, USA.

Yoshikawa, T., Furuhashi, T., & Uchikawa, Y. (1996). "Emergence of Effective Fuzzy Rules for Controlling Mobile Robots using DNA Coding Method", Proceedings of ICEC'96, Nagoya, Japan, pp. 581–586.

Yuille, A., Cohen, D., and Hallinan, P. (1989). "Feature Extraction from Faces using Deformable Templates", Proceedings of IEEE Conference on Computer Vision and Templates, pp. 104–109.

Zadeh, L. A. (1965). "Fuzzy Sets", *Journal of Information and Control*, Vol. 8, pp. 338–353.

Zadeh, L. A. (1971a). "Similarity Relations and Fuzzy Ordering", *Journal of Information Sciences*, Vol. 3, pp. 177–206.

Zadeh, L. A. (1971b). "Quantitative Fuzzy Semantics", *Journal of Information Sciences*, 3, pp. 159–176.

Zadeh, L. A. (1973). "Outline of a New Approach to the Analysis of Complex Systems and Decision Processes", *IEEE Transactions on Systems, Man and Cybernetics*, Vol. 3, pp. 28–44.

Zadeh, L. A. (1975). "The Concept of a Linguistic Variable and its Application to Approximate Reasoning–1", *Journal of Information Sciences*, Vol. 8, pp. 199–249.

Index

activation function
  hyperbolic tangent function 59
  identity function 59
  logistic function 59
adaptive learning rate 64
adaptive network-based fuzzy inference
  system 75
antecedent 19
architecture
  for face recognition 190
  for fingerprint recognition 4, 207,
    208, 212, 214, 216, 250
  for human recognition 5, 241, 250
  for voice recognition 5, 223, 234,
    238, 239
architecture optimization 157
attractor 104

backpropagation learning rule 56, 60
binary fuzzy relations 14, 15
biometric
  information 1, 4, 5, 31, 42, 85, 94,
    115, 116, 118, 119, 131, 132, 134,
    141, 155, 158, 165, 169, 188, 208,
    214, 215, 223, 230, 241, 246, 247,
    250
  measures 5, 64, 118–120, 142, 164,
    176, 177, 187, 200, 214, 241, 245,
    246, 248
  modalities 246, 248
  techniques 1, 4, 5, 16, 31, 32, 56, 58,
    70, 72, 80, 94, 98, 122, 129, 131,
    132, 139, 142, 143, 160, 164, 168,

169, 179, 187–189, 191, 207, 208,
    223, 225, 226, 241, 242, 244, 246
  templates 188, 227, 244, 246–248
bisector 20

centroid 44–46, 237, 252
cepstral coefficients 225, 226
chromosome 133, 134, 140–144, 149,
    151–153, 155, 156, 158, 161–163,
    166, 214, 215
chromosome representation 140, 166,
    215
cluster analysis 4, 169
cluster centers 86–88, 94, 95, 171, 173,
    174, 177–179
clustering validity 175, 176, 184
coding 131, 133, 135, 140
committees of networks 110
competitive learning 85, 87–89, 94,
    107, 113, 182
competitive networks 191, 193, 196
complement 10, 38, 41, 99
concentration 16, 165
consequents 2, 19, 33, 43, 45, 47
containment 10
convergence theorem 173
core 9, 10
correlation-based method 207
crossover 9, 13, 133–135, 137, 143, 146,
    149–151, 153, 157, 159, 162, 166,
    215

defuzzification 19–21, 25, 26, 42–44,
    46, 48